International Energy Outlook 2014

Contents

Overview 1
   Prices 1
   Demand 1
Reference case 3
   World petroleum and other liquid fuels consumption 4
      OECD 5
      Non-OECD 6
   World petroleum and other liquid fuels supplies 7
      OPEC crude and lease condensate supply 7
      Non-OPEC crude and lease condensate supply 9
         Americas 9
         Other non-OPEC crude and lease condensate supply 12
   Other liquids supply 13
High Oil Price case 14
Low Oil Price case 15
Energy reform in Mexico 16
Endnotes 18
Data sources 20
Appendix A 23
Appendix B 35
Appendix C 47

Tables

Table 1. North Sea Brent crude oil spot prices in three cases, 2010-40 2
Table 2. World gross domestic product by OECD and non-OECD in three oil price cases, 1990-2040 4
Table 3. World petroleum and other liquid fuels consumption by region, Reference case, 1980-2040 5
Table 4. World liquid fuels production in the Reference case, 2010-40 10
Table 5. World other liquid fuels production by fuel type, 2010-40 14

Figures

Figure 1. North Sea Brent crude oil spot prices in three cases, 1990-2040 2
Figure 2. World tight oil production in the Reference case, 2010 and 2040 2
Figure 3. Liquid fuels consumption and production in three cases, 2040 3
Figure 4. Liquid fuels supply and demand and North Sea Brent crude oil equilibrium prices in three cases, 2040 3
Figure 5. OECD and Non-OECD petroleum and other liquid fuels consumption, Reference case, 1990-2040 4
Figure 6. Non-OECD petroleum and other liquid fuels consumption by region, Reference case, 1990-2040 4
Figure 7. Petroleum and other liquid fuels consumption in China and the United States, Reference case, 1990-2040 6
Figure 8. Middle East liquid fuels consumption by end-use sector, 2010-40 6
Figure 9. Petroleum and other liquid fuels production by region and type in the Reference case, 2000-2040 7
Figure 10. OPEC crude and lease condensate production by region in the Reference case, 2010 and 2040 7
Figure 11. Non-OPEC crude and lease condensate production, 2010 and 2040 9
Figure 12. Brazil and Argentina crude and lease condensate production, 2010-40 11
Figure 13. Non-OECD Europe and Eurasia crude and lease condensate production, 2010-40 12
Figure 14. World other liquid fuels production by source, 2010 and 2040 13
Figure 15. World petroleum and other liquid fuels consumption by country grouping in three cases, 2010 and 2040 ... 15
Figure 16. World petroleum and other liquid fuels production in three cases, 2010 and 2040 15
Figure 17. Mexico petroleum and other liquid fuels production in IEO2013 and IEO2014, 2010-40 17

Overview

World markets for petroleum and other liquid fuels have entered a period of dynamic change—in both supply and demand. Potential new supplies of oil from tight and shale resources have raised optimism for significant new sources of global liquids. The potential for growth in demand for liquid fuels is focused on the emerging economies of China, India, and the Middle East, while liquid fuels demand in the United States, Europe, and other regions with well-established oil markets seems to have peaked. After a long period of sustained high oil prices, improvements in conservation and efficiency have reduced or slowed the growth of liquid fuels use among mature oil consumers. The changes in the overall market environment have led the U.S. Energy Information Administration (EIA) to focus on reassessing long-term trends in liquid fuels markets for the 2014 edition of its *International Energy Outlook* (IEO2014).

IEO2014 projections of future liquids balances include two broad categories: crude and lease condensate and other liquid fuels. Crude and lease condensate includes tight oil, shale oil, extra-heavy crude oil, field condensate, and bitumen (i.e., oil sands, either diluted or upgraded). Other liquids refer to natural gas plant liquids (NGPL), biofuels (including biomass-to-liquids [BTL]), gas-to-liquids (GTL), coal-to-liquids (CTL), kerogen (i.e., oil shale), and refinery gain.

After the oil crises of the 1970s and 1980s, much of the debate about world oil markets centered on the limitations of supply. Energy security was (and remains) a major concern, with large resource deposits located in and controlled by members of the Organization of the Petroleum Exporting Countries (OPEC). In addition, strong increases in demand for oil and a limited supply response to rising prices in the mid-2000s led to increasingly vocal concerns about resource depletion. More recently, with higher sustained world oil prices—by historic measures—and advances in extraction technologies, growing supplies of tight oil and shale oil in the United States have brought new resources to market, beginning in North America and, eventually, in other parts of the world. There is also hope that recent legislative changes in Mexico will reverse that country's recent trend of slowly declining oil production. Outside North America, the potential for large production increases in Brazil, Argentina, and elsewhere could help ensure the availability of liquid fuels supplies for many years.

Prices

The benchmark oil price used in IEO2014 is based on spot prices for North Sea Brent crude oil, which is an international standard for light sweet crude oil. The West Texas Intermediate (WTI) spot price and the North Sea Brent price diverged in 2011 to a high of around $30 per barrel, with Brent the more expensive oil. The spread eventually decreased in 2013 and 2014 as a result of new crude pipeline construction and crude pipeline flow reversals in the United States. EIA expects the WTI-Brent discount to continue to decrease over time and will continue to report WTI prices (a critical reference point for the value of growing production in the U.S. Midcontinent), as well as the imported refiner acquisition cost (IRAC). The recent decision by the U.S. Commerce Department's Bureau of Industry and Security to allow exports of some lease condensates after processing also has the potential to further reduce the spread between the Brent price and the price of domestic production streams.

Since July 2012, North Sea Brent prices have generally remained in the range of $100 to $115 (nominal dollars) per barrel. Growing liquids supplies in North America—especially from the United States and also from Canada—have brought more than 4 million barrels per day (MMbbl/d) of additional liquids supplies to market since 2008. However, that increase has largely been offset by supply disruptions in other oil-producing regions, notably in North Africa and the Middle East. EIA estimates that unplanned crude oil production outages have averaged 2.7 MMbbl/d over the past two years and generally have trended upward, from 1.8 MMbbl/d in May 2012 to about 3.5 MMbbl/d in May 2014 [7]. OPEC member countries Libya and Iran and non-OPEC countries South Sudan and Syria have accounted for a sizeable portion of the unplanned outages. It is difficult to predict when the supplies may return, given the significant geopolitical difficulties faced by these producers. This adds considerable uncertainty to the mid-term and long-term projections.

Demand

Demand has also played a role in keeping world oil prices largely stable for the past few years. In countries outside the Organization for Economic Cooperation and Development (non-OECD countries), demand growth has moderated as key economies, including China, India, and Brazil, have seen slower economic growth and correspondingly slower growth in liquids demand compared with the past two decades. Liquids consumption among OECD countries, which reached its peak at 50 MMbbl/d in 2005, has generally been trending downward since that time, reflecting both slow economic growth and growing energy efficiency in the transportation sector.

Key influences on consumption and production are price trends and the reactions of consumers and producers to those trends, which in turn influence future prices. EIA has developed three price cases to examine a range of potential interactions of supply, demand, and prices in world liquids markets: the Reference case and alternative Low Oil Price and High Oil Price cases.

While the three oil price cases represent a wide range of uncertainty in future markets, they do not capture all possible outcomes (Figure 1 and Table 1). Because EIA's oil price paths represent market equilibrium between supply and demand, they do not show the price volatility[1] that occurs over days, months, or years. As a frame of reference, over the past two decades oil price volatility

---

[1] Historical price volatility is a measure of the magnitude of daily price movements in percentage terms over a specified period. It is measured by calculating the standard deviation (movement from the average) over the past 30 days of the daily percentage changes in price, and then annualizing the percentage. Market participants use measures of volatility to gauge uncertainty in factors that influence oil markets, including supply, demand, and geopolitical or macroeconomic events.

within single years has averaged about 30%. Although that level of volatility could continue, the alternative oil price cases in IEO2014 assume smaller near-term price variation than in previous IEOs, because larger near-term price swings are expected to lead to market changes in supply or demand that would dampen price volatility.

In the IEO2014 Reference case, world oil prices fall from $113 per barrel (2012 dollars) in 2011 to $92 per barrel in 2017, then rise steadily to $141 per barrel in 2040. Worldwide consumption of petroleum and other liquid fuels rises from 87 MMbbl/d in 2010 to 98 MMbbl/d in 2020 and 119 MMbbl/d in 2040. Compared with last year's IEO, IEO2014 incorporates a larger increase in production from non-OPEC producers, particularly the United States. The largest new supplies of tight oil come from the United States, although a few other countries, including Canada, Mexico, Russia, Argentina, and China, also begin producing substantial volumes of tight oil in the Reference case (Figure 2). In addition, IEO2014 assumes that the OPEC countries will choose to maintain their market share of world liquid fuels production, and as a result, they will schedule investments in incremental production capacity so that total OPEC liquid fuels production represents between 39% and 44% of the world total throughout the projection.

Production of other liquids increases by an average of 1.7% per year in the Reference case—almost twice as fast as crude and lease condensate production. The growth in other liquid supplies is attributed to byproducts of natural gas production (in the case of NGPL) and government policies aimed at increasing the use of alternative liquid fuels in the transportation sector. Other liquid supplies account for between 14% and 17% of total liquid supplies throughout the projection period.

The IEO2014 Low and High Oil Price cases were developed by adjusting four key factors:

- Energy demand growth, especially in the non-OECD countries, which accounts for much of the uncertainty about future demand growth
- OPEC investment and production decisions
- Non-OPEC crude and lease condensate supply
- Other liquid fuels supply.

In 2040, both the Low Oil Price and High Oil Price cases have higher liquids demand (123 MMbbl/d and 122 MMbbl/d, respectively) than the Reference case (119 MMbbl/d) (Figure 3). The three cases provide an assessment of alternative views on the future courses of both production and consumption of liquids, as summarized in the supply and demand curves shown in Figure 4 for 2040. The Low Oil Price case assumes slower economic growth in combination with lower cost of producing petroleum and other liquids than in the Reference case. The Low Oil Price case demand curve is shifted downward relative to the Reference case curve, indicating less demand for liquid fuels at a given oil price. The Low Oil Price case supply curve is also shifted down, reflecting greater supply at a given oil price. In contrast, the High Oil Price case assumes faster economic growth and a higher cost of producing petroleum and other liquids than in the Reference case. In the High Oil Price case, the demand and supply curves are shifted upward relative to the Reference case, indicating greater demand and less supply at a given price.

Table 1. North Sea Brent crude oil spot prices in three cases, 2010-40 (2012 dollars per barrel)

Year	Reference	Low Oil Price	High Oil Price
2010	83	83	83
2020	97	69	150
2025	109	70	159
2030	119	72	174
2035	130	73	188
2040	141	75	204

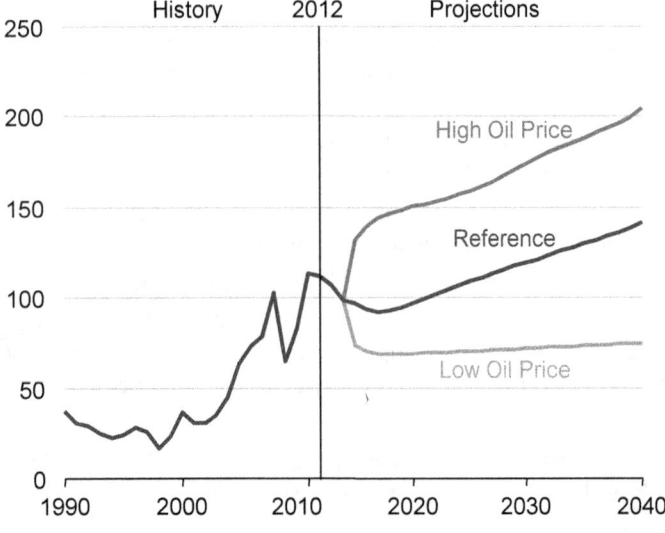

Figure 1. North Sea Brent crude oil spot prices in three cases, 1990-2040 (2012 dollars per barrel)

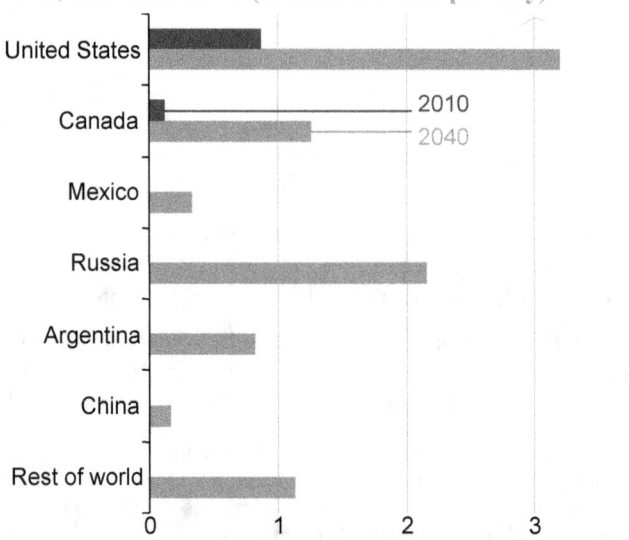

Figure 2. World tight oil production in the Reference case, 2010 and 2040 (million barrels per day)

In the Low Oil Price case, crude oil prices fall to $70 per barrel (2012 dollars) in 2016, then remain below $70 per barrel through 2023 and below $75 per barrel through 2040. Gross domestic product (GDP) growth in the non-OECD countries averages 4.2% per year from 2010 to 2040 in the Low Oil Price case, compared with Reference case growth of 4.6% per year. The combination of lower economic activity and lower prices in the Low Oil Price case results in non-OECD liquid fuels consumption in 2040 that is close to that in the Reference case, at 73.7 MMbbl/d and 74.7 MMbbl/d, respectively. In contrast, in the OECD regions, where economic growth is the same in the Low Oil Price and Reference cases, lower prices encourage consumers to use more liquid fuels.

On the supply side, OPEC countries increase their output above the Reference case level in the Low Oil Price case, achieving a 53% share of total world liquid fuels production by 2040. However, liquids production in the non-OPEC countries is lower in the Low Oil Price case than in the Reference case, because their more expensive resources cannot be brought to market economically.

In the High Oil Price case, oil prices are about $204 per barrel (2012 dollars) in 2040. GDP growth in the non-OECD regions averages 5.0% per year from 2010 to 2040, compared with Reference case growth of 4.6% per year. With higher economic activity, non-OECD liquid fuels consumption reaches 80.1 MMbbl/d in 2040, 5.4 MMbbl/d higher than in the Reference case (see Figure 3). The increase in non-OECD liquid fuels consumption is partially offset by a decline in OECD liquid fuels consumption, as consumers improve efficiency or switch to less expensive fuels when possible.

On the supply side, liquid fuels production in the OPEC countries is lower in the High Oil Price case than in the Reference case, and their market share of total petroleum and other liquid fuels production declines to between 34% and 37%. However, high world oil prices allow non-OPEC countries to increase production from more costly resources, and their crude and lease condensate production in the High Oil Price case increases to 61.3 MMbbl/d in 2040, 8.4 MMbbl/d higher than in the Reference case. The economics of other liquid fuels also benefit from higher prices, with production increasing to 22.9 MMbbl/d in 2040, 2.7 MMbbl/d higher than in the Reference case. Across the three price cases, OPEC's petroleum and other liquid fuels production decreases as oil prices rise from the Low Oil Price case to the High Oil Price case. Alternatively, non-OPEC production of both petroleum and other liquid fuels increases as oil prices increase from the Low Oil Price case to the High Oil Price case.

The following discussion reviews the three price cases, their assumptions, and implied trends, along with the potential effect of each set of factors on the future evolution of liquids markets. The following sections discuss the adjustments made in IEO2014. Each price case represents one of potentially many combinations of supply and demand that would result in the same price path. EIA does not assign probabilities to any of the oil price cases.

Reference case

In the IEO2014 Reference case, world oil prices trend downward from current rates to about $97 per barrel in 2020.[2] After 2020, prices trend upward, to $141 per barrel in 2040—considerably lower than the $165 per barrel projected in IEO2013. Consumption of world petroleum and other liquid fuels rises from 87 MMbbl/d in 2010 to 98 MMbbl/d in 2020 and 119 MMbbl/d in 2040. Crude and lease condensate supplies from OPEC and non-OPEC sources rise from 74.9 MMbbl/d in 2010 to 99.1 MMbbl/d in 2040, an increase of 24.2 MMbbl/d. Over the projection, OPEC member countries account for 14.2 MMbbl/d, and non-OPEC countries account for 10.0 MMbbl/d, of the total increase in crude and lease condensate production. Production of other liquid fuels rises from 12.3 MMbbl/d in 2010 to 20.3 MMbbl/d in 2040.

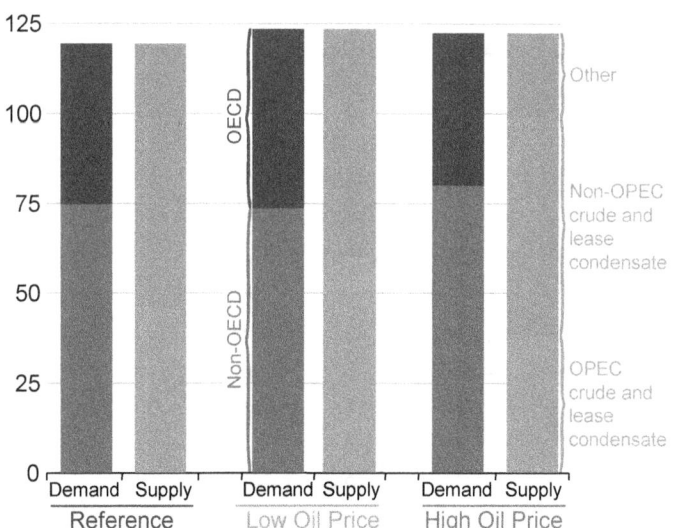

Figure 3. Liquid fuels consumption and production in three cases, 2040 (million barrels per day)

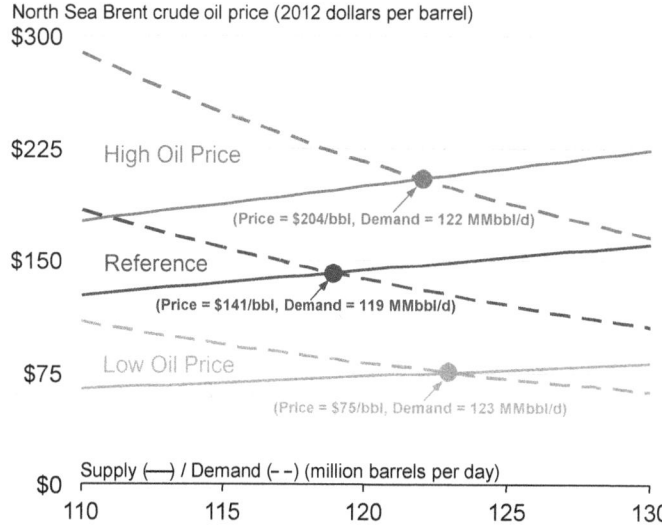

Figure 4. Liquid fuels supply and demand (million barrels per day) and North Sea Brent crude oil equilibrium prices (2012 dollars per barrel) in three cases, 2040

[2] Unless otherwise noted, all prices are reported in inflation-adjusted 2012 U.S. dollars.

World petroleum and other liquid fuels consumption

In the IEO2014 Reference case, world liquid fuels consumption increases by more than one-third (33 MMbbl/d), from 87 MMbbl/d in 2010 to 119 MMbbl/d in 2040. Rising prices for liquid fuels improve the cost competitiveness of other fuels, leading many users of liquid fuels outside the transportation and industrial sectors to switch to other sources of energy when possible. The transportation and industrial sectors account for 92% of global liquid fuels demand in 2040, whereas in every other end-use sector the consumption of liquid fuels decreases on a worldwide basis over the projection period.

Economic growth is among the most important factors to be considered in projecting changes in world energy consumption. In IEO2014, assumptions about regional economic growth—measured in terms of real GDP in 2005 U.S. dollars at purchasing power parity rates—underlie the projections of regional energy demand. World economic growth averaged less than 3% in both 2012 and 2013. In the IEO2014 Reference case, global GDP is expected to rise at an average annual rate of 3.5% from 2010 to 2040. The fastest economic growth is projected for the non-OECD region, with GDP increasing by an average of 4.6% per year from 2010 to 2040. In contrast, GDP in the OECD region rises by only 2.1% per year over the same period (Table 2).

Non-OECD regions account for virtually all of the increase in demand for petroleum and other liquid fuels in the Reference case (Figure 5). In particular, non-OECD Asia and the Middle East account for 85% of the total increase in world liquid fuels consumption (Figure 6). Fast-paced economic expansion among the non-OECD regions drives the increase in demand for liquid fuels, as strong growth in income per capita raises demand for personal transportation and freight transport, as well as demand for energy in the industrial sector.

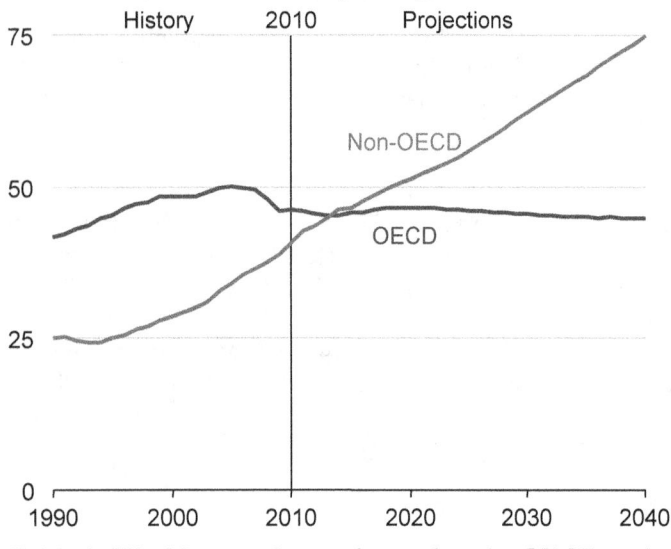

Figure 5. OECD and Non-OECD petroleum and other liquid fuels consumption, Reference case, 1990-2040 (million barrels per day)

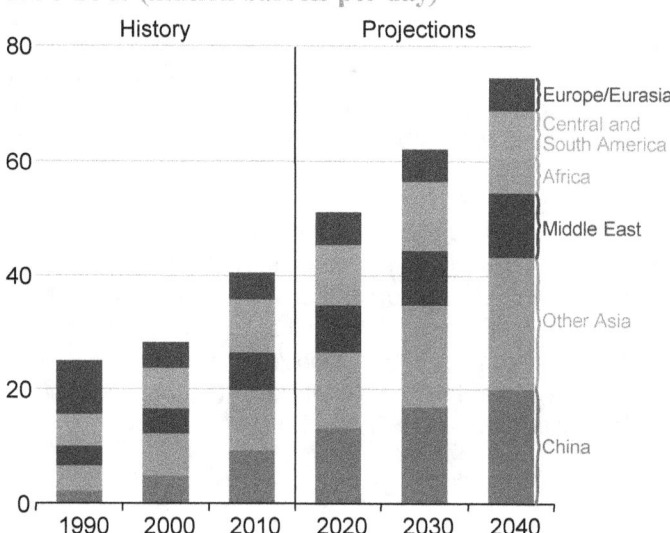

Figure 6. Non-OECD petroleum and other liquid fuels consumption by region, Reference case, 1990-2040 (million barrels per day)

Table 2. World gross domestic product by OECD and non-OECD in three oil price cases, 1990-2040 (billion 2005 dollars, purchasing power parity)

	History			Projections					Average annual percent change, 2010-40
	1990	2009	2010	2020	2025	2030	2035	2040	
Reference case									
OECD	23,529	35,551	36,609	45,711	50,965	56,358	62,143	68,357	2.1
Non-OECD	14,416	31,640	33,889	56,432	72,255	90,672	111,157	132,123	4.6
World	**37,945**	**67,192**	**70,498**	**102,142**	**123,220**	**147,030**	**173,300**	**200,479**	**3.5**
High Oil Price case									
OECD	23,529	35,551	36,609	45,643	51,126	56,434	62,431	68,950	2.1
Non-OECD	14,416	31,640	33,889	56,952	73,998	94,949	119,767	146,431	5.0
World	**37,945**	**67,192**	**70,498**	**102,595**	**125,124**	**151,383**	**182,197**	**215,381**	**3.8**
Low Oil Price case									
OECD	23,529	35,551	36,609	45,831	51,029	56,461	62,204	68,623	2.1
Non-OECD	14,416	31,640	33,889	55,231	69,188	84,442	100,481	117,012	4.2
World	**37,945**	**67,192**	**70,498**	**101,062**	**120,217**	**140,903**	**162,685**	**185,635**	**3.3**

OECD liquid fuels consumption declines by 0.1% per year, as the mature economies react to sustained high fuel prices. Strong efficiency gains (especially in personal transportation) and conservation reduce OECD demand for liquid fuels. While technology efficiency improvements and fuel-switching opportunities are also available to non-OECD consumers, the sheer scale of growth in demand for transportation services in relatively underdeveloped transportation networks overwhelms the advantages of those improvements.

OECD

Consumption of petroleum and other liquid fuels in the OECD countries remains relatively flat or declines in the Reference case (Table 3). Total OECD liquid fuels use decreases by 1.4 MMbbl/d, from 46.0 MMbbl/d in 2010 to 44.7 MMbbl/d in 2040. In much of the OECD, relatively stable economic growth and static or declining population levels contribute to lower levels of liquid fuels consumption. In addition, many OECD governments have adopted policies that mandate improvements in the efficiency of motor vehicles, and consumers turn to more fuel-efficient transportation choices in the face of sustained high oil prices.

The United States currently is the OECD's largest consumer of liquid fuels, and it remains so through 2040. The use of liquid fuels in the U.S. transportation sector declines over the projection period, as a result of significantly lower energy use by light-duty vehicles; however, the decline is moderated somewhat by increased energy use for heavy-duty vehicles, aircraft, and marine vessels. Over the course of the projection, increases in vehicle fuel economy offset growth in transportation activity. In addition, industrial sector demand for liquid fuels in the United States grows over the projection, mainly because of increased use of hydrocarbon gas liquids (HGL)—primarily ethane and propane—as a feedstock in the bulk chemicals industry. Total liquid fuels consumption in the United States rises from 18.9 MMbbl/d in 2010 to 19.2 MMbbl/d in 2020, then declines to 18.4 MMbbl/d in 2040.

New motor vehicles in Canada and Mexico are likely to show fuel efficiency gains similar to those in the United States, as motor vehicle markets across the continent tend to be interconnected. In Canada, the result of the fuel efficiency gains is relatively flat consumption of petroleum and other liquid fuels, between 2.1 and 2.3 MMbbl/d throughout the projection. In Mexico and Chile combined, liquid fuels consumption rises by 0.7% per year—the highest growth rate among the OECD regions. Despite improvements in vehicle fuel efficiency, the use of liquid fuels increases in Mexico and Chile, particularly for transportation services (as the infrastructure is still relatively underdeveloped) and for industrial production as demand grows.

In OECD Europe, consumption of liquid fuels declines as a result of improvements in energy efficiency. In addition to improvements in motor vehicle fuel efficiency, most nations in OECD Europe have high taxes on motor fuels, well-established public transportation systems, and declining or slowly growing populations, all of which slow the growth of transportation energy use. In 2040, liquid fuels consumption in OECD Europe totals 14.0 MMbbl/d, or 0.8 MMbbl/d lower than the 2010 level of 14.8 MMbbl/d.

Petroleum and other liquid fuels consumption in OECD Asia is expected generally to decline over the long term, from 7.7 MMbbl/d in 2010 to 7.2 MMbbl/d in 2040. In the short term, however, the region's liquid fuels consumption rises, largely because of increased fuel use in Japan. Demand for liquid fuels in Japan increased after the March 2011 earthquake and tsunami that severely damaged nuclear reactors at Fukushima Daiichi and subsequently led to the shutdown of all the country's nuclear power reactors by May 2012. To compensate for the loss of nuclear generation, Japan turned, in part, to oil-fired generation to meet demand for electricity in the short term. Consumption of petroleum and other liquid fuels for power generation increased by 22% from 2011 to 2012. As damaged coal-fired generators and some nuclear reactors return to service, Japan's use of liquid fuels for power generation is expected to return to more typical levels, and the country's overall trend of decreasing petroleum and other liquid fuels consumption is expected to resume in the medium to long term.

Table 3. World petroleum and other liquid fuels consumption by region, Reference case, 1980-2040 (million barrels per day)

Region	1980	1990	2000	2010	2020	2030	2040	Average annual percent change 1980-2010	2010-40
OECD	41.6	41.6	48.3	46.0	46.4	45.3	44.7	0.3	-0.1
Americas	20.3	20.4	24.0	23.5	24.3	23.6	23.5	0.5	0.0
Europe	15.2	13.9	15.6	14.8	14.1	14.0	14.0	-0.1	-0.2
Asia	6.2	7.2	8.7	7.7	8.0	7.7	7.2	0.7	-0.2
Non-OECD	21.5	24.9	28.5	40.7	51.2	62.1	74.7	2.2	2.0
Europe/Eurasia	10.1	9.3	4.3	4.8	5.5	5.6	5.6	-2.4	0.5
Asia	4.5	6.6	12.2	19.8	26.5	34.8	43.2	5.0	2.6
Middle East	1.9	3.3	4.5	6.7	8.4	9.6	11.1	4.3	1.7
Africa	1.5	2.1	2.5	3.4	3.9	4.8	6.2	2.8	2.0
Central/South America	3.5	3.6	5.0	6.0	6.9	7.4	8.6	1.8	1.2
World	63.1	66.5	76.8	86.8	97.6	107.4	119.4	1.1	1.1

Outside of Japan, the other countries of OECD Asia—South Korea, Australia, and New Zealand—experience modest growth in liquids fuels use, attributed to expanding transportation and industrial activity. In Australia and New Zealand, expected population growth rates also contribute to a rise in demand for liquid fuels.

Non-OECD

The non-OECD share of world liquid fuels consumption grows substantially over time, from 47% in 2010 to 63% in 2040. Non-OECD Asia shows the largest growth in liquid fuels consumption worldwide in the Reference case, at 23.4 MMbbl/d from 2010 to 2040, with China accounting for 10.7 MMbbl/d of the total increase. As China's economy moves from dependence on energy-intensive industrial manufacturing to services, the transportation sector becomes the most significant source of growth in liquid fuels use, and the country's liquid fuels consumption more than doubles from its 2010 level. In the Reference case, China replaces the United States as the world's largest consumer of liquid fuels by 2035 (Figure 7).

For India, IEO2014 projects a lower GDP growth rate than in the IEO2013 outlook. India's economic growth rate has declined in recent years, to 4.1% in 2013 from 7% to 8% or more over the past decade. Slowing industrial activity and sluggish service sector activity, combined with the lack of structural reform, have contributed to the lower expectations for economic growth [2]. In addition, the country's continued efforts to reduce subsidies on petroleum products are expected to temper demand for liquid fuels. In the IEO2014 Reference case, consumption of petroleum and other liquid fuels in India rises from 3.3 MMbbl/d in 2010 to 6.8 MMbbl/d in 2040, about 1.4 MMbbl/d lower than projected in IEO2013.

Liquid fuels demand in the Middle East grows substantially in the IEO2014 Reference case, by 4.4 MMbbl/d from 2010 to 2040, as a result of strong population growth rates, second only to those in Africa, and rising incomes. Liquids-intensive industrial demand also plays a major role in the region, with consumption in the chemical sector leading the growth of industrial demand. Delays in petroleum subsidy reforms (outside of Iran) and strong growth of income per capita support significant expansion of transportation sector demand for liquid fuels in the region (Figure 8). In the later years of the projection, it is likely that some subsidy reform will occur, and the resulting higher prices will begin to slow the growth in demand for liquid fuels.

In the Middle East, demand for liquid fuels in the electric power sector declines from 2010 to 2040 in the Reference case. Many of the countries in the region that produce liquid fuels increasingly turn to lower-cost natural gas and, to a lesser extent, nuclear and renewable fuels, in an effort to increase the volumes of petroleum available for export. The timing of the Middle East shift from reliance on liquid fuels for power generation remains uncertain, however, as the region faces delays in improving the infrastructure, and there are limits on the supply of alternative fuels for power generation. For instance, Saudi Arabia has been unable to meet rapid growth in electricity demand with power generated from domestic natural gas and has had to import fuel oil for power generation [3].

As in the Middle East, growing populations and economies in African countries increase the demand for liquid fuels for both transportation and industrial uses. In the IEO2014 Reference case, Africa's consumption of petroleum and other liquid fuels increases by 2.8 MMbbl/d from 2010 to 2040. Real GDP in Africa increases by 4.8% per year from 2010 to 2040. With an expected favorable investment environment and relative political stability in the long term, growing consumer demand drives demand for consumer goods and services and for liquid fuels, particularly for personal transportation and freight services. In the IEO2014 Reference case projection, more than 80% of the total increase in liquid fuels use in Africa is in the transportation sector.

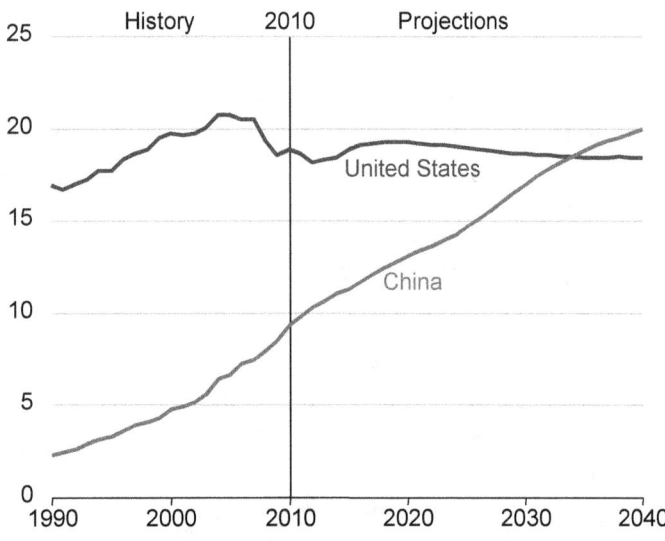

Figure 7. Petroleum and other liquid fuels consumption in China and the United States, Reference case, 1990-2040 (million barrels per day)

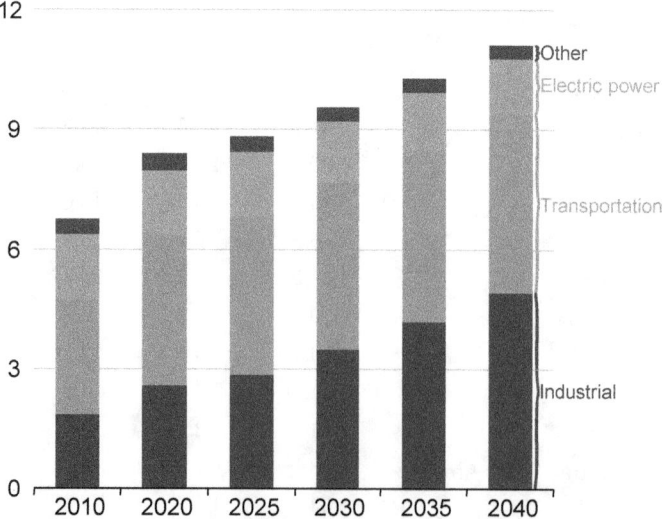

Figure 8. Middle East liquid fuels consumption by end-use sector, 2010-40 (million barrels per day)

In Brazil and the other nations of Central and South America, consumption of liquid fuels increases by 2.6 MMbbl/d, from 6.0 MMbbl/d in 2010 to 8.6 MMbbl/d in 2040. Some regional economies—notably Brazil, Colombia, and Peru—are expected to continue to experience long-term economic expansion that will support growing demand for liquid fuels, primarily for transportation uses but also in the industrial sector. Brazil, with the region's largest economy, accounts for more than half of the regional growth in liquid fuels demand in the Reference case. Regional economies that are less financially secure than Brazil's, including Venezuela and Argentina, will have a more difficult time sustaining economic growth. Fuel subsidies in Venezuela, in particular, are very costly, although it is difficult to anticipate when the Venezuelan government might be able to reduce subsidies.

In the countries of non-OECD Europe and Eurasia, demand for liquid fuels is projected to rise moderately from 2010 to 2020 before reaching a plateau. Russia—the largest economy in the region—currently accounts for the largest share of the region's consumption of liquid fuels, but its consumption increases more slowly than in other parts of non-OECD Europe and Eurasia as a result of major efficiency improvements in its energy-intensive industrial sector. In addition, demand for liquid fuels in Russia's residential and commercial sectors is projected to slow as fuel subsidies for people living in areas with high heating requirements are reduced.

World petroleum and other liquid fuels supplies

The Reference case assumes that OPEC will maintain a cohesive policy limiting supply growth, rather than maximizing total annual revenues. It also assumes that no geopolitical events will cause prolonged supply shocks in the OPEC countries that could further limit production growth. Accordingly, world oil prices trend downward, from $113 per barrel in 2011 to about $92 in 2017, and then increase steadily to $141 per barrel in 2040.

OPEC crude and lease condensate[3] supply

The IEO2014 Reference case assumes that OPEC producers will invest in incremental production capacity that enables them to increase crude and lease condensate production by 14.2 MMbbl/d from 2010 to 2040 (Figure 9) and to account for 41% to 47% of total crude and lease condensate production worldwide over the course of the projection. The Middle East OPEC member countries, which accounted for 68% of total OPEC crude and lease condensate production in 2010 (Figure 10), are projected to increase their crude and lease condensate production by 12.8 MMbbl/d in the Reference case, accounting for 90% of the total growth in OPEC crude and lease condensate production from 2010 to 2040.

Saudi Arabia, Iran, and Iraq combined have a large share of the world's oil reserves and resources that are relatively inexpensive to produce. Saudi Arabia, which has for some time been the only holder of substantial spare oil production capacity, has played a critical role as the major swing supplier in response to disruptions in other supply sources and economic fluctuations that affect oil demand. Both Iraq and Iran have the reserves needed to raise their capacity and production well above current levels if they can successfully address some of the internal and external *above-ground constraints*[4] that have kept their respective oil sectors from realizing their potential for more than 30 years. There is considerable uncertainty in projecting the extent to which these countries will be able to overcome the difficulties that impede supply growth.

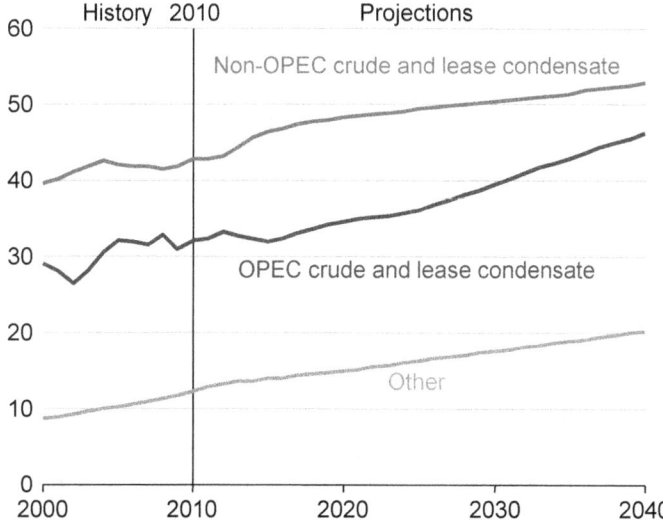

Figure 9. Petroleum and other liquid fuels production by region and type in the Reference case, 2000-2040 (million barrels per day)

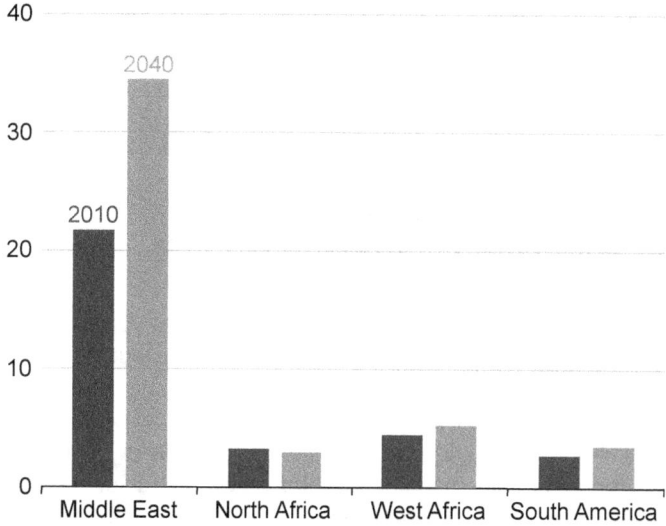

Figure 10. OPEC crude and lease condensate production by region in the Reference case, 2010 and 2040 (million barrels per day)

[3] Crude and lease condensate includes tight oil, shale oil, extra-heavy oil, field condensate, and bitumen (i.e., oil sands, either diluted or upgraded).
[4] *Above-ground constraints* are those nongeological factors that could affect supply, including but not limited to government policies that limit access to resources; conflict; terrorist activity; lack of access to technology; price constraints on the economic development of resources; labor shortages; materials shortages; weather; environmental protection actions; and short- and long-term geopolitical considerations.

Regardless of the uncertainties in oil supply projections, producers in the OPEC Middle East region are likely to continue playing a key role in balancing global demand and supply. As a result, their output levels may be negatively correlated, with higher realizations of capacity and production in one country reducing the amounts of capacity and production in other countries that are needed to balance global markets. Future developments, including production from tight oil resources—which is a topic widely discussed in the international oil community—have significant potential to affect the world's reliance on OPEC liquids supplies and the behavior of key Middle East OPEC producers over the next several decades.

Saudi Arabia's oil revenue traditionally exceeds the amount required to fund its government expenditures, enabling it to vary production levels in response to global supply or demand developments over the past 25 years without significant concern about the short-term revenue implications of such actions. More recently, social and economic programs funded by the Saudi government have expanded substantially. While Saudi Arabia maintains large financial reserves, revenue needs may become a more important consideration as the government considers its future responses to a situation of persistent high growth in supply from other key OPEC or non-OPEC producers, or a sustained downturn in demand.

In 2013, Iraq confirmed that it had revised its official oil production target down to 9 MMbbl/d by 2017, from its previous target of 12 MMbbl/d [4]. This level still represents a very ambitious target, and Iraqi officials have also stated that sustained output of 5 MMbbl/d to 6 MMbbl/d would be sufficient to allow the country to meet its revenue requirements. Even before the most recent turmoil in Iraq, it was unlikely that the country would achieve its new production target (which would also exceed the amount of global incremental liquid fuels production needed to meet projected global demand growth to 2017 in the Reference case). The June 2014 attacks by the insurgent Islamic State of Iraq and Syria (ISIS) adds considerable uncertainty to the prospects for Iraq's future stability and governance, as well as its oil production. No matter how the latest event is resolved, political instability and delays in establishing a regulatory framework, along with infrastructure limitations, are likely to continue to impede output growth. In addition, terrorism, a poor investment climate, and other problems will be limiting factors for Iraq's oil production over the projection period. If the country's profound difficulties could be overcome, major improvements in production and export infrastructure would make it possible for Iraq to sustain high production growth rates through 2040.

Iran's liquid fuels production, which reached a peak of 6.1 MMbbl/d in 1974, has been well below that level since 1979 [5]. After averaging an estimated 4.0 MMbbl/d from 2001 to 2010, Iran's production has continued declining, to an estimated 3.2 MMbbl/d in 2013. A series of international sanctions targeting Iran's oil sector have led foreign companies to cancel a number of new projects and upgrades of existing projects. Iran faces continued depletion of its production capacity, as its fields have relatively high natural decline rates (between 8% and 13% per year). Additional factors hampering investment include unfavorable foreign investment requirements, underinvestment, and gaps in professional expertise and technology for certain projects. U.S. sanctions on financial institutions that handle payments made for oil exports from Iran, coupled with actions by the European Union to cease imports from Iran and prevent it from accessing insurance from European Union companies for its oil shipments, led to a further reduction in Iran's oil exports in 2012.

In November 2013, the P5+1 countries (China, France, Russia, the United Kingdom, and the United States, plus Germany) agreed to a temporary suspension of a number of European Union and U.S. sanctions in exchange for Iran's agreement to cease development of its nuclear program and limit uranium stockpiles to 5% enrichment [6]. The temporary agreement is aimed at giving the parties additional time to negotiate a longer-term arrangement. Oil-related sanctions remain in place, but petrochemical exports are allowed as part of the agreement [7]. Successful negotiation of a long-term agreement between the parties remains highly uncertain, and it is likely to take many years for Iran to return oil production to the levels achieved in the 1970s.

The remaining Middle East OPEC producers are expected to make smaller, but important, contributions to supply in the future. For example, nearly all of Kuwait's current reserves and production are in mature fields, but prospects could improve with the success of Project Kuwait, a plan first proposed in 1998 to attract foreign participation and to increase oil production capacity from four northern oil fields: Raudhatain, Sabriya, al-Ratqa, and Abdali. The four fields contain a mix of heavy and light oil resources. Additionally, it may be possible for Kuwait to boost oil production from the partitioned neutral zone (PNZ) that the country shares with Saudi Arabia, which could hold as much as 5 billion barrels of oil [8]. Qatar's liquids production is poised to increase over the projection period through the application of GTL technology, which produces liquid fuels such as low-sulfur diesel and naphtha from natural gas.

West African OPEC crude and lease condensate production increases to 5.3 MMbbl/d in 2040 in the Reference case, from 4.4 MMbbl/d in 2010. Nigeria has increased its output from deepwater fields in recent years, but onshore production has declined as infrastructure constraints and incidents of oil theft and attacks on pipelines have curbed production growth and are expected to continue in the near- to mid-term. Angola is expanding its offshore deepwater production and, as relative geopolitical stability improves, is likely to develop onshore exploration and production areas as well [9]. To meet the government's goal of maintaining oil production at around 2 MMbbl/d, state-owned Sonangol plans to make substantial exploration and development investments in deepwater and ultra-deepwater areas of its Congo Fan region, as well as to develop presalt resources in the Kwanza and Benguela basins [10]. IHS Energy has estimated that around $30 billion is expected to be invested in 12 deepwater developments between 2013 and 2020 [11].

In the IEO2014 Reference case, **crude and lease condensate production in OPEC's North African member countries**—Libya and Algeria—is projected to decline from 3.2 MMbbl/d in 2010 to 3.0 MMbbl/d in 2040. The potential for growth in Libya's production is high, but the country has been unable to stabilize production amid social and political unrest. After the 2011 overthrow of the Muammar al-Gaddafi regime, Libya's crude and lease condensate production returned to pre-revolution levels of about 1.6 MMbbl/d in October 2012, but with ongoing political unrest and mechanical problems, production levels have continued to decline, to less than 0.5 MMbbl/d [12]. Until a permanent government is in place, it will be difficult to improve conditions sufficiently to attract the foreign investment needed to repair and improve Libya's production infrastructure. As a result, the country's prospects for increased production are unlikely to improve substantially for several years.

North African OPEC member Algeria has also encountered difficulties in improving its petroleum production. State-owned Sonatrach was forced to delay its target date to raise crude oil production to 2.0 MMbbl/d by 2010, with actual production at around 1.1 MMbbl/d in 2013 [13]. Exploration investment in Algeria's oil sector has declined since 2006, as a result of amendments to the country's hydrocarbon law that were unfavorable to foreign investment. The law was amended again in 2013 in an attempt to attract more foreign investment, but positive results are not anticipated until well into the midterm, and perhaps later.

South American OPEC crude and lease condensate production increases by 0.9% per year in the Reference case, to 3.5 MMbbl/d in 2040. Venezuela is the dominant producer in South American OPEC, which also includes Ecuador. There are abundant proved reserves of extra-heavy oil in Venezuela's Orinoco belt, but bringing those resources to market will require substantial investment, which may be difficult for Venezuela to attract without policy and political changes. In March 2014, state-owned oil company Petróleos de Venezuela, S.A. (PDVSA) announced its plans to increase extra-heavy oil production from Orinoco by 2 MMbbl/d over a six-year period [14]. This is an ambitious target and one that will require more favorable terms to attract foreign investment and participation. Social unrest in the form of antigovernment protest, extremely high inflation, and a rising homicide rate are undermining popular support for the government. Given the importance of petroleum revenues to the government, there is some urgency to increase PDVSA's oil production as quickly as possible.

Non-OPEC crude and lease condensate supply

In the IEO2014 Reference case, non-OPEC crude and lease condensate production increases steadily from 42.9 MMbbl/d in 2010 to 48.3 MMbbl/d in 2020 and 52.9 MMbbl/d in 2040, as rising world oil prices attract investment in areas previously considered uneconomical (Table 4). Sustained high oil prices encourage producers in non-OPEC nations to continue investment in new production capacity and to apply advanced technologies, such as enhanced oil recovery (EOR), horizontal drilling, and hydraulic fracturing, to increase recovery rates from existing fields. The average cost per barrel of non-OPEC oil production rises as production volumes increase, and the rising costs eventually dampen further production growth.

In the early years of the IEO2014 Reference case projection, non-OPEC crude and lease condensate production grows by 5.4 MMbbl/d to 48.3 MMbbl/d in 2020, primarily as a result of increased production from U.S. tight oil formations (see Figure 9). After 2020, production growth continues at a slower pace, adding another 4.6 MMbbl/d to net petroleum production by 2040, with production from new wells increasing slightly faster than the decline in production from existing wells.

Almost all of the net increase in non-OPEC crude and lease condensate production is attributed to only five countries: Canada, Brazil, the United States, Kazakhstan, and Russia (Figure 11). One potential wildcard for future increases in non-OPEC petroleum is Mexico, where recent changes to the country's constitutional prohibition against foreign participation in the oil and natural gas sectors could profoundly improve prospects for long-term production (see box on page 16). EIA's projection incorporates a conservative increase in Mexico's long-term production profile, reversing the trend noted in prior IEOs that had projected declining production. Still, there is considerable upside potential for improved production from existing onshore reservoirs, as well as deepwater Gulf of Mexico production and tight oil formations that extend from the U.S. border.

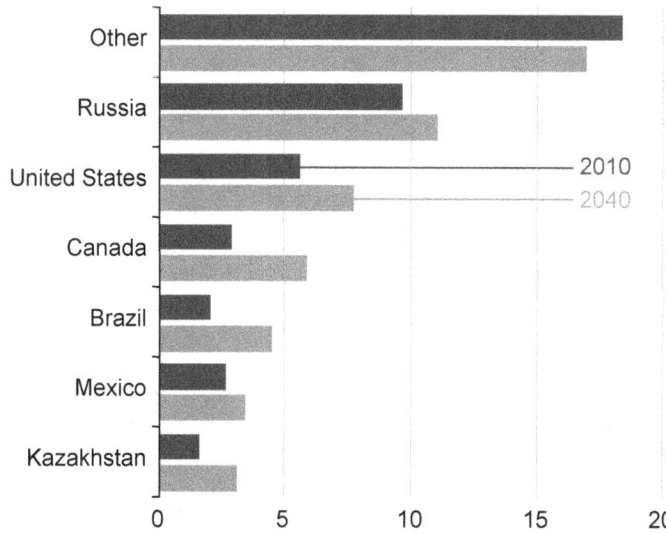

Figure 11. Non-OPEC crude and lease condensate production, 2010 and 2040 (million barrels per day)

Americas

Canada. In the IEO2014 Reference case, Canada has the largest increase in production among the non-OPEC countries, with crude and lease condensate production increasing by 3.0 MMbbl/d from 2010 to 2040. Much of the projected growth results from a strong increase in bitumen production from Alberta's oil sands, along with the growth of tight oil production. Increased production of bitumen and tight oil offsets declining production of crude and lease condensate. In 2012, production from oil sands accounted for nearly 60% of Canada's oil output, a proportion that has increased steadily. Alberta accounted for about 75% of

Canada's oil production in 2012, according to data from Canada's National Energy Board [15]. In addition, there are a number of promising tight oil formations in the Western Canada Sedimentary Basin (WCSB) provinces of Alberta, Manitoba, Saskatchewan, and British Columbia. Canada's light oil production has already reversed the long-term decline in the WCSB, and the trend is likely to continue in the future [16].

United States. The United States is expected to be another important source of non-OPEC growth in the Reference case, with U.S. crude and lease condensate production growing from 2010 through 2019 before peaking at 9.9 MMbbl/d—about 4.3 MMbbl/d above the 2010 total and equal to the 1970 historical U.S. high. The growth in Lower 48 states onshore crude oil production is primarily a result of continued development of tight oil resources in the Bakken, Eagle Ford, and Permian Basin formations. U.S. tight oil production increases to 4.8 MMbbl/d in 2021 (compared to 0.9 MMbbl/d in 2010) and then declines to about 3.2 MMbbl/d in 2040 (2.3 MMbbl/d higher than the 2010 total) as high-productivity areas (called sweet spots) are depleted. There is considerable uncertainty about the expected peak level of tight oil production because exploration, appraisal, and development programs are expanding operator knowledge about producing reservoirs. This knowledge could result in the identification of additional tight oil resources or the ability to exploit resources more fully.

U.S. oil production from the offshore Lower 48 states varies between 1.4 MMbbl/d and 2.0 MMbbl/d over the projection period. Toward the end of the period the pace of exploration and production activity quickens. As a result, new large development projects associated predominantly with discoveries in the deepwater and ultradeepwater portions of the Gulf of Mexico are brought on stream. In addition, new offshore oil production from the Alaska North Slope partially offsets the decline in production from onshore North Slope fields.

Table 4. World liquid fuels production in the Reference case, 2010-40 (million barrels per day)

Source	2010	2020	2025	2030	2035	2040	Average annual percent change 2010-40
OPEC							
Crude and lease condensate[a]	32.0	34.4	36.1	39.5	42.9	46.2	1.2
Natural gas plant liquids	3.3	4.0	4.2	4.5	4.9	5.4	1.7
Biofuels[b]	0.0	0.0	0.0	0.0	0.0	0.0	--
Coal-to-liquids	0.0	0.0	0.0	0.0	0.0	0.0	--
Gas-to-liquids	0.0	0.3	0.3	0.4	0.4	0.4	14.1
Kerogen	0.0	0.0	0.0	0.0	0.0	0.0	--
Refinery gain	0.0	0.0	0.0	0.0	0.1	0.1	0.9
Total OPEC	**35.4**	**38.7**	**40.7**	**44.4**	**48.2**	**52.1**	**1.3**
Non-OPEC[c]							
Crude and lease condensate[a]	42.9	48.3	49.4	50.4	51.4	52.9	0.7
Natural gas plant liquids	5.1	5.9	6.4	6.7	7.1	7.3	1.2
Biofuels[b]	1.3	1.8	2.1	2.4	2.7	3.0	2.7
Coal-to-liquids	0.2	0.3	0.5	0.7	0.9	1.1	6.2
Gas-to-liquids	0.1	0.1	0.1	0.2	0.2	0.2	3.9
Kerogen	0.0	0.0	0.0	0.0	0.0	0.0	0.6
Refinery gain	2.3	2.5	2.6	2.7	2.8	2.9	0.8
Total Non-OPEC	**51.9**	**58.9**	**61.1**	**63.1**	**64.9**	**67.2**	**0.9**
World							
Crude and lease condensate[a]	74.9	82.7	85.5	89.9	94.3	99.1	0.9
Natural gas plant liquids	8.4	9.9	10.6	11.2	11.9	12.7	1.4
Biofuels[b]	1.3	1.8	2.1	2.4	2.7	3.0	2.7
Coal-to-liquids	0.2	0.3	0.5	0.7	0.9	1.1	6.2
Gas-to-liquids	0.1	0.3	0.4	0.5	0.6	0.6	7.6
Kerogen	0.0	0.0	0.0	0.0	0.0	0.0	0.6
Refinery gain	2.3	2.5	2.6	2.7	2.8	2.9	0.8
Total World	**87.2**	**97.6**	**101.8**	**107.4**	**113.1**	**119.4**	**1.1**

[a]Crude and lease condensate includes tight oil, shale oil, extra-heavy oil, and bitumen.
[b]Ethanol volumes are reported on a gasoline-equivalent basis.
[c]Includes some U.S. petroleum product stock withdrawals, domestic sources of blending components, other hydrocarbons, and ethers.

The IEO2014 Reference case reflects continued growth in U.S. tight oil production. However, growth potential and sustainability of domestic crude oil production hinge around uncertainties in key assumptions, such as well production decline, lifespan, drainage areas, geologic extent, and technological improvement—both in areas currently being drilled and those yet to be drilled. As a result, High and Low Oil and Gas Resource cases were developed for the United States in the *Annual Energy Outlook 2014*[5] to examine the effects of alternative resource and technology assumptions on U.S. production.

The projected trends in U.S. oil production vary significantly in the alternative resource cases, and those trends hold important implications for the United States. The High Oil and Gas Resource case assumes a broad-based future increase in crude oil and natural gas resources, not limited to production of oil and natural gas in tight sands and shales. However, optimism about supply increases has been buoyed by recent advances in the production of crude oil and natural gas from tight and shale formations. With the adjusted resource and technology advance assumptions in the High Oil and Gas Resource case, U.S. crude oil production (including lease condensate) continues to increase, to 13.4 MMbbl/d in 2036, and remains above 13 MMbbl/d through 2040. The Low Oil and Gas Resource case reflects uncertainties about U.S. resources of tight oil, shale crude oil, and natural gas that lower domestic production relative to the Reference case. In this case, production reaches 9.1 MMbbl/d in 2017 before falling to 6.6 MMbbl/d in 2040.

Brazil. Brazil's crude and lease condensate production increases by 2.4 MMbbl/d from 2010 to 2040 in the Reference case (Figure 12). Some of the world's largest oil discoveries in recent years have been in Brazil's offshore presalt basins. Along with the potential to increase the country's oil production significantly, the presalt areas are estimated to contain sizable natural gas reserves. State-owned Petrobras has set a crude oil production target of 2.75 MMbbl/d by 2017 [17]. Achieving that goal will depend on the successful development of Brazil's significant offshore presalt resources, which are made producible because of relatively high oil prices and improvements in extraction technology and techniques for very deep water.

The IEO2014 Reference case anticipates substantial increases in production of Brazil's presalt resources, but future large-scale development of those resources will require the participation of companies with the technical resources and capital needed to produce them. The large deepwater presalt discoveries in Brazil are estimated to include as much as 28 billion barrels of oil equivalent. Petrobras first announced a large oil discovery in the presalt layer of the Santos Basin in November 2007, with initial estimates of between 5 and 8 billion barrels of oil equivalent [18]. Commercial production at Sapinhoá, the first presalt field, began in January 2013, and output has exceeded the original production estimates [19].

Although it is technically challenging and costly, oil production from Brazil's presalt layers in the Santos and Campos basins has substantial potential for robust growth. It is clear that there will have to be favorable terms to attract foreign investment from companies with technology that can help develop the resources. The Brazilian government has ruled that Petrobras will be the sole operator of each presalt production sharing agreement and will hold a minimum 30% stake in all presalt projects, which may be a drawback for foreign investors. Another potential impediment to foreign investment is that Brazil's local-content rules require 37% of the labor and materials associated with the exploration phase of presalt development to be supplied domestically, with the percentage increasing over time to 59% for wells that begin producing oil after 2021 [20]. Brazil's 12th licensing round—and the first involving presalt licensing—took place in October 2013 for the Santos Basin's Libra field and received only one bid, from a consortium of Petrobras, Shell, Total, China National Offshore Oil Corporation (CNOOC), and China National Petroleum Company (CNPC) [21].

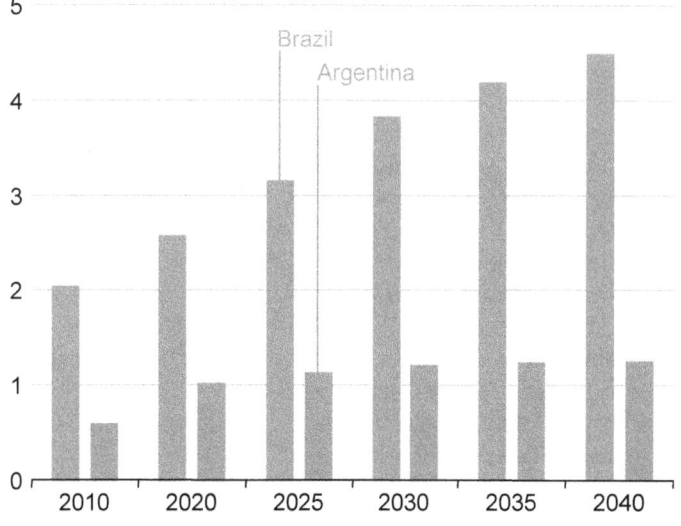

Figure 12. Brazil and Argentina crude and lease condensate production, 2010-40 (million barrels per day)

Argentina. Argentina currently produces about 0.5 MMbbl/d of crude oil. While production of crude oil and lease condensate is likely to decline over the projection period, the country does have the potential to produce tight oil in volumes similar to those from the U.S. Eagle Ford and Bakken formations. The 2013 report, *Technically Recoverable Shale Oil and Shale Gas Resources: An Assessment of 137 Shale Formations in 41 Countries Outside the United States* [22], estimates Argentina's technically recoverable shale oil resources at 27 billion barrels, the second largest shale oil resource in the Americas after the United States. A number of major international energy companies are actively exploring the potential of Argentina's shale resources. As of June 2013, more than 50 test wells had been drilled in Los Molles and Vaca Muerta in the Neuquen Basin [23]. There is a potential for investment problems because of the nationalization of assets and inflation and also because the existing regulatory framework discourages foreign participation. The IEO2014 Reference case assumes that tight oil will be Argentina's main source of new oil production and will more than offset declines in production of crude oil and

---

[5]For a full discussion of the High and Low Oil and Gas Resource cases, see "IF2. U.S. tight oil production: Alternative supply projections and an overview of EIA's analysis of well-level data aggregated to the county level" in EIA's *Annual Energy Outlook 2014*, DOE/EIA-0383(2014) (Washington, DC, April 2014), pp. IF10-IF14, http://www.eia.gov/forecasts/aeo/.

lease condensate. Progress on development of the Vaca Muerta shale play will be a good measure of the potential for successful exploitation of Argentina's shale oil resources.

Other non-OPEC crude and lease condensate supply

Non-OECD Europe and Eurasia. In the IEO2014 Reference case, Kazakhstan and Russia account for virtually all crude and lease condensate production growth in non-OECD Europe and Eurasia (Figure 13). Crude and lease condensate production in Kazakhstan grows by 1.6 MMbbl/d, from 1.6 MMbbl/d in 2010 to 3.1 MMbbl/d in 2040. While Kazakhstan is one of the largest sources of additional non-OPEC crude oil production in EIA's projection, the size of the increase has been lowered since the IEO2013 was published. Performance has been disappointing at the supergiant offshore Kashagan field, where the project took more than a decade to bring production online (in September 2013) only to have it shut in soon after it started because of a gas leak [24]. Further, Kazakhstan's Minister of Oil and Gas announced in April 2014 that production at Kashagan may not resume before the end of the year [25]. Despite substantial foreign investment, the difficulties of working in the offshore Caspian terrain have tempered prospects for growth.

Outside of Kashagan, Kazakhstan has three other major onshore fields—Tengiz, Uzen, and Karachaganak—in the western part of the country [26]. Petroleum production from Tengiz alone doubled between 2004 and 2008, reaching 0.54 MMbbl/d in September 2008 when its full expansion was completed. Although crude oil and lease condensate production at Tengiz has declined somewhat since 2008, to 0.47 MMbbl/d in 2012, the Tengizchevriol consortium (led by Chevron) expects to increase petroleum production further, to as much as 1 MMbbl/d [27]. Uzen and Karachaganak have smaller reserves than Tengiz but are expected to contribute to an increase in crude and lease condensate production in the next several years.

Russia is the region's top oil producer, and with rich resources and new investment in exploration it is likely to remain an important liquid fuels producer in the future. At present, much of Russia's oil production comes from fields in the country's West Siberian Basin; however, interest is shifting toward undeveloped resources in East Siberia, the Russian Arctic, the northern Caspian Sea, and Sakhalin Island [28]. Russia's crude and lease condensate production declined precipitously after the fall of the Soviet Union, bottoming out at 5.9 MMbbl/d in 1996. Since then, the country has been able to increase crude and lease condensate production to levels achieved during the Soviet era, with production of 10.0 MMbbl/d in 2012.

Russia is concerned about ensuring that it can maintain its position as one of the world's largest crude and lease condensate producers and recognizes that it will not be possible without substantial investment in exploration. In 2013, exploration expenditures by Russian oil companies exceeded $8.4 billion, with another $1 billion contributed by the Russian government [29]. If successful, the government expects that about 3 MMbbl/d of onshore crude oil production could be added to the country's total production by 2035, with additional production from the Arctic Pechora Sea and offshore resources. In addition, there have been efforts to create incentives to develop Russia's large tight oil resources. In September 2013, the Russian government introduced new tax rules that reduced the mineral extraction tax to zero for certain tight oil formations, including the large Bazhenov formation in Western Siberia and other tight oil plays in southern Russia and in the Volga-Urals region [30]. The IEO2014 Reference case assumes that some of the planned capacity will be successful, and that Russia will be able to increase crude and lease condensate production from 9.7 MMbbl/d in 2010 to 11.1 MMbbl/d in 2040.

Russia's annexation of Crimea in March 2014 and its role in the unrest in eastern Ukraine have raised tensions between Russia and the West. So far, Western sanctions that have targeted specific Russian citizens seem to have had little impact on the Russian energy sector. In fact, agreements between international oil companies (for example, Total and BP) were reached with Russian energy companies at the St. Petersburg International Economic Forum in May 2014, for exploration projects in Russia's Volga-Urals and Western Siberia [31]. A new wider and more stringent round of sanctions was announced in late July 2014, specifically targeting Russia's energy sector. The sanctions, which aim to prevent Russia from obtaining the technology it requires to develop new streams of oil production, particularly in existing tight oil formations or in the Arctic, could have a profound effect on the country's future production potential [32]. IEO2014 does not incorporate any effect from the Russia-Ukraine events in the projection, but it is possible that further escalation of tensions could make investors nervous about entering into new activities or expanding their current investments.

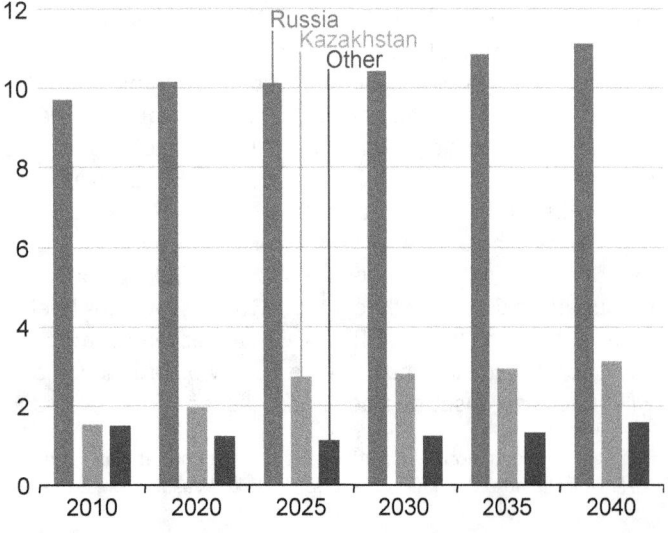

Figure 13. Non-OECD Europe and Eurasia crude and lease condensate production, 2010-40 (million barrels per day)

North Sea. The North Sea continental shelf contains significant oil reserves and is the largest source of oil production in OECD Europe. According to international agreements, five countries (Denmark, Germany, Netherlands, Norway, and the United

Kingdom) can award licenses for crude oil production in the area. Several production streams from the North Sea constitute the Brent international benchmark for oil prices. However, since reaching peak production of 5.8 MMbbl/d in 1996, crude and lease condensate output from the North Sea has been declining slowly, to 2.4 MMbbl/d in 2013.

In the Reference case, declines in North Sea crude and lease condensate production continue, averaging 2.6% per year, with production falling from 3.1 MMbbl/d in 2010 to 1.4 MMbbl/d in 2040. The largest decline is in the United Kingdom's production, as a result of depleting reserves and an aging oil infrastructure [33]. Norway may have some potential to offset the decline in North Sea production. There is continued strong interest in exploration investment in the Norwegian Continental Shelf, which includes parts of the North Sea, as well as the Norwegian Sea and the Barents Sea [34]. The Barents Sea in particular is relatively unexplored and may offer the greatest potential to reverse the decline in Norwegian oil production. However, lack of infrastructure and recent cost and tax increases have resulted in delays, including the development of the Johan Castberg project, which was discovered in 2011 and includes recoverable reserves estimated between 400 and 600 million barrels [35].

Non-OECD Asia. With the exceptions of China and India, few countries in non-OECD Asia are able to increase their crude and lease condensate production. In the IEO2014 Reference case, China's crude and lease condensate production increases from 4.1 MMbbl/d in 2010 to 4.7 MMbbl/d in 2025, before declining back to 4.1 MMbbl/d in 2040. China's largest oil fields, located in the northeast and north central regions of the country, represent the backbone of the country's domestic production; however, the oil fields are mature and prone to declining production, although the use of enhanced oil recovery (EOR) techniques has been somewhat successful in slowing the decline rates. Much of the projected increase results from investments by Chinese national oil companies, such as CNPC and CNOOC; the latter is responsible for offshore oil exploration and production in locations such as the South China Sea.

India's crude and lease condensate production increases by 0.3% per year in the IEO2014 Reference case, from 0.7 MMbbl/d in 2010 to 0.8 MMbbl/d in 2040. While the Mumbai Offshore and Assam-Arakan basins have been India's main sources of crude oil supplies historically, large oil discoveries have been made recently in the Barmer Basin in Rajasthan and the Krishna-Godavari Basin offshore of Andhra Pradesh [36]. The new discoveries demonstrate India's resolve to diversify its oil supplies and to allow continued steady growth in crude oil production through the projection period.

Outside of China and India, there are few prospects for increasing crude and lease condensate production in non-OECD Asia. In the IEO2014 Reference case, non-OECD Asia's oil production in countries that include Vietnam, Indonesia, and Thailand declines by an average of 1.3% per year through 2040 because of aging petroleum fields and a lack of substantial new discoveries.

Other liquids supply

Other liquid resources—including natural gas plant liquids (NGPL), biofuels, coal-to-liquids (CTL), gas-to-liquids (GTL), kerogen (oil shale), and refinery gain—currently supply a relatively small portion of total world petroleum and other liquid fuels, accounting for about 14% of the total in 2010. However, they are expected to grow in importance. In the IEO2014 Reference case, the other liquids share of the world's total liquids supply rises to 17% in 2040 (see Figure 9).

In a change from previous IEO reports, NGPL have been classified as other liquid fuels. NGPL are those hydrocarbons in natural gas that are separated as liquids at natural gas processing, fractionating, and cycling plants. Growing production of wet natural gas and lighter crude oil has focused attention on natural gas liquids, and EIA has developed and adopted the neutral term *hydrocarbon gas liquids* to equate supply (i.e., NGPL and liquefied refinery gases) with markets (i.e., natural gas liquids and refinery olefins). The tables included in IEO2014 incorporate this revised definition.

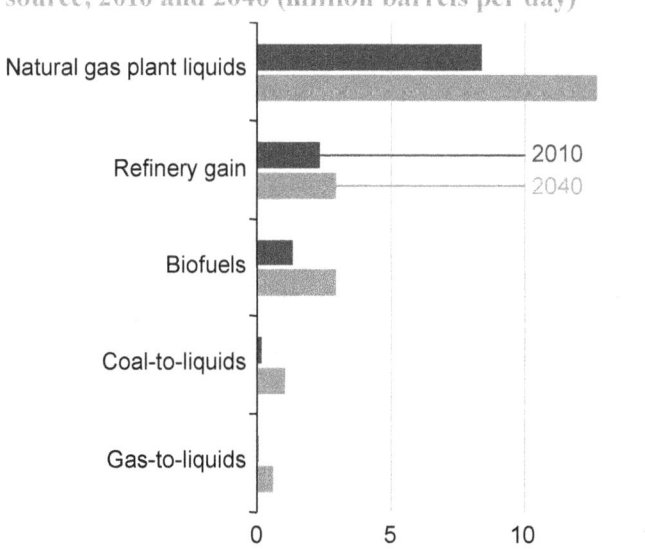

Figure 14. World other liquid fuels production by source, 2010 and 2040 (million barrels per day)

NGPL are the largest component of the other liquids, accounting for 68% of the total in 2010 (Figure 14). The increase in NGPL production is directly correlated to the increase in natural gas production. In contrast, increased production of the remaining other liquids (primarily biofuels, CTL, and GTL) is in response to policies that encourage growth in the expansion of these liquids with available domestic resources, such as coal and crops. In the IEO2014 Reference case, sustained high oil prices make the development of the non-NGPL other liquids more attractive. In addition, biofuels development also relies heavily on country-specific programs or mandates. Combined, the remaining, non-NGPL other liquid fuels grow at more than twice the rate of NGPL over the projection period.

Non-OPEC production of other liquid fuels grows from 9.0 MMbbl/d in 2010 to 14.4 MMbbl/d in 2040 (Table 5), accounting for two-thirds of the growth in world other liquids. Non-OPEC producers add 2.1 MMbbl/d of NGPL over the projection—about 70% of which come from the United States and Russia alone. Other important increases in alternative liquids supply include an additional 0.5 MMbbl/d of biofuels from Brazil and an additional 0.3 MMbbl/d of biofuels and 0.6 MMbbl/d of CTL from China.

Similar to the non-OPEC producers, NGPL is the main constituent of OPEC other supplies, accounting for more than 90% of the total even in 2040. NGPL production in OPEC countries rises by 2.2 MMbbl/d, from 3.3 MMbbl/d in 2010 to 5.4 MMbbl/d in 2040. The only other form of OPEC alternative supplies is GTL, primarily from Qatar. OPEC GTL production increases to 0.4 MMbbl/d in 2040.

High Oil Price case

In the High Oil Price case, the assumed higher costs of crude and lease condensate production result in lower production than in the Reference case. In addition, the higher prices result in increased development of liquid supplies from emerging sources, including tight oil and bitumen, which have higher production costs. World oil prices (in 2012 dollars) are $150 per barrel in 2020 and $204 per barrel in 2040.

Higher oil prices depress demand globally through 2033, but stronger economic growth in non-OECD countries compared with the Reference case leads to higher demand for liquid fuels in those countries. In the High Oil Price case, non-OECD GDP grows at an average annual rate of 5.0% per year from 2010 to 2040, as compared with 4.6% per year in the Reference case. In 2040, non-OECD consumption of petroleum and other liquid fuels totals 80.1 MMbbl/d, or 5.4 MMbbl/d higher than in the Reference case. The increase in non-OECD demand for liquid fuels is only partially offset by a decline in OECD demand as consumers improve efficiency or switch to less-expensive fuels when possible (Figure 15). OECD consumption of liquid fuels declines by 4.0 MMbbl/d, from 46.0 MMbbl/d in 2010 to 42.0 MMbbl/d in 2040, as consumers implement efficiency and conservation measures and switch to alternative fuels.

On the supply side, the high oil prices allow non-OPEC countries to increase production from more costly resources. Non-OPEC crude and lease condensate production increases initially in the High Oil Price case at about the same rate as in the Reference case, because access to existing resources is limited, and discovery rates are lower. Eventually, however, non-OPEC crude and

Table 5. World other liquids by fuel type, 2010-40 (million barrels per day)

Source	2010	2020	2025	2030	2035	2040	Average annual percent change 2010-40
OPEC							
NGPL	3.27	3.97	4.25	4.51	4.89	5.43	1.7
Biofuels[a]	0.00	0.00	0.00	0.00	0.00	0.00	--
Coal-to-liquids	0.00	0.00	0.00	0.00	0.00	0.00	--
Gas-to-liquids	0.01	0.27	0.30	0.35	0.40	0.42	14.1
Kerogen	0.00	0.00	0.00	0.00	0.00	0.00	--
Refinery gain	0.04	0.04	0.05	0.05	0.05	0.05	0.9
Total OPEC	**3.32**	**4.28**	**4.60**	**4.91**	**5.33**	**5.91**	**1.9**
Non-OPEC							
NGPL	5.13	5.92	6.40	6.69	6.96	7.27	1.2
Biofuels[a]	1.33	1.83	2.11	2.42	2.70	2.97	2.7
Coal-to-liquids	0.17	0.33	0.51	0.69	0.87	1.05	6.2
Gas-to-liquids	0.06	0.06	0.13	0.16	0.17	0.19	3.9
Kerogen	0.01	0.01	0.01	0.01	0.01	0.01	0.6
Refinery gain	2.30	2.48	2.56	2.68	2.78	2.89	0.8
Total Non-OPEC	**9.00**	**10.62**	**11.71**	**12.65**	**13.63**	**14.37**	**1.6**
World							
NGPL	8.40	9.89	10.64	11.20	11.85	12.71	1.4
Biofuels[a]	1.33	1.83	2.11	2.42	2.70	2.97	2.7
Coal-to-liquids	0.17	0.33	0.51	0.69	0.87	1.05	6.2
Gas-to-liquids	0.07	0.33	0.43	0.51	0.57	0.61	7.6
Kerogen	0.01	0.01	0.01	0.01	0.01	0.01	0.6
Refinery gain	2.34	2.52	2.61	2.73	2.83	2.94	0.8
Total World	**12.32**	**14.90**	**16.31**	**17.56**	**18.83**	**20.27**	**1.7**

[a]Ethanol volumes are reported on a motor-gasoline-equivalent basis.

lease condensate production grows to 61.3 MMbbl/d in 2040, or 8.4 MMbbl/d higher than in the Reference case (Figure 16). The economics of other liquids also benefit from higher prices. Non-OPEC production of other liquid fuels increases to 16.8 MMbbl/d in 2040 in the High Oil Price case, nearly 3 MMbbl/d higher than in the Reference case. Non-OPEC production of NGPL grows to 7.8 MMbbl/d in 2040, only about 0.5 MMbbl/d higher than in the Reference case, but higher prices for North Sea Brent crude oil lead to significant increases in non-OPEC production of biofuels, CTL, and GTL as compared with the Reference case. In 2040, non-OPEC other supplies (excluding NGPL) are 2.1 MMbbl/d higher in the High Oil Price case than in the Reference case.

The High Oil Price case assumes that OPEC successfully limits crude and lease condensation production to maintain higher prices. As a result, the OPEC market share of world petroleum and other liquids production never exceeds the high of 41% it reached in 2012, declining to as low as 34% after 2020 and eventually climbing back to 36% in 2040. OPEC petroleum and other liquids production rises from 35.4 MMbbl/d in 2010 to 43.7 MMbbl/d in 2040, which is 8.4 MMbbl/d lower than in the Reference case.

Low Oil Price case

In the Low Oil Price case, world oil prices fall to $70 per barrel (2012 dollars) in 2016, remain below $70 per barrel through 2023, and do not rise above $75 per barrel through 2040. The low price results not only from an assumption of lower marginal production costs but also from assumed lower demand for liquid fuels in the non-OECD nations, especially China and India, as compared with the Reference case. OPEC production of crude oil and lease condensate increases throughout the projection, displacing more expensive petroleum and other liquid fuels production. Resources that are more costly to produce are not economically viable in the Low Oil Price case.

The IEO2014 Low Oil Price case assumes slower economic growth in the non-OECD countries than in the Reference case, with combined GDP in the non-OECD region increasing by 4.2% per year from 2010 to 2040, as compared with 4.6% per year in the Reference case (see Table 2). However, with the impact of lower economic activity offset by lower oil prices, non-OECD consumption of liquid fuels rises from 40.7 MMbbl/d in 2010 to 51.6 MMbbl/d in 2020 and 73.7 MMbbl/d in 2040, which is only 1 MMbbl/d below the level in the Reference case (see Figure 15). In contrast, economic growth in the OECD as a whole is the same in the Low Oil Price case as in the Reference case—2.1% per year from 2010 to 2040—and the lower prices encourage consumers to use more liquid fuels. In the Low Oil Price case, OECD liquid fuels consumption rises from 46.0 MMbbl/d in 2010 to 47.8 MMbbl/d in 2020 and 49.6 MMbbl/d in 2040. In contrast, in the Reference case, OECD liquid fuels consumption totals 46.4 MMbbl/d in 2020 and 44.7 MMbbl/d in 2040.

On the supply side, OPEC's market share of crude and lease condensate production remains around 42% through 2016, then rises to 46% in 2020 and to 57% in 2040. OPEC is assumed to be less successful in restricting production to try to keep prices high in the Low Oil Price case relative to the Reference case, and as a result OPEC crude and lease condensate production increases by 28.2 MMbbl/d, from 32.0 MMbbl/d in 2010 to 60.2 MMbbl/d in 2040. In contrast, because North Sea Brent prices are lower than in the Reference case, non-OPEC crude and lease condensate production levels increase by only about 2 MMbbl/d, to 44.7 MMbbl/d in 2040, or 8.2 MMbbl/d lower than 2040 production in the Reference case. With higher average costs for resource development in the non-OPEC countries, the North Sea Brent crude oil price in the Low Oil Price case is not sufficient to make undeveloped fields economically viable. Production of other liquid fuels that are typically more expensive to produce also grows more slowly than in the Reference case. Total other liquid fuels production increases from 12.3 MMbbl/d in 2010 to 18.4 MMbbl/d in 2040, or 1.9 MMbbl/d lower than projected in the Reference case.

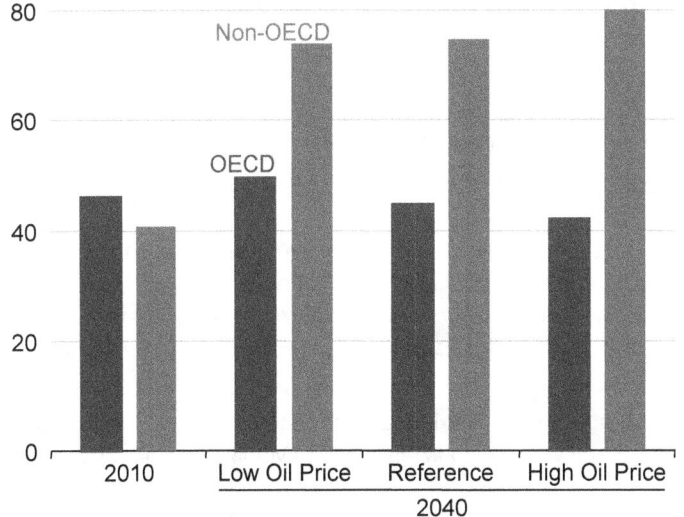

Figure 15. World petroleum and other liquid fuels consumption by country grouping in three cases, 2010 and 2040 (million barrels per day)

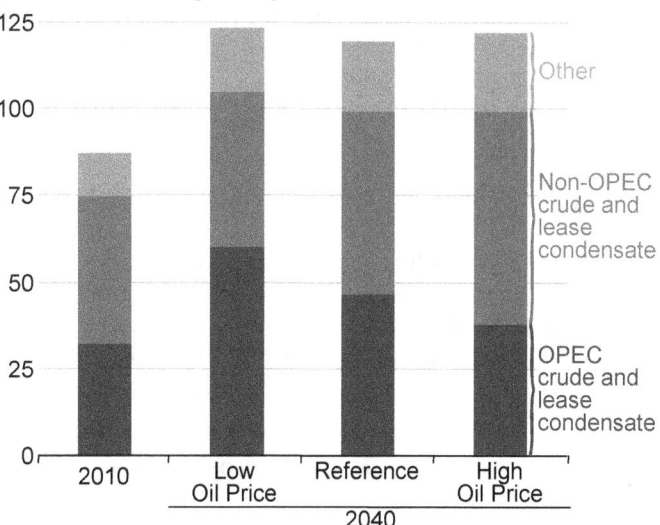

Figure 16. World petroleum and other liquid fuels production in three cases, 2010 and 2040 (million barrels per day)

Energy Reform in Mexico

In December 2013, Mexico's government approved constitutional amendments that would alter the 1938 nationalization of the energy sector, effectively ending the 75-year monopoly of state-owned Petróleos Mexicanos (Pemex) and allowing for more foreign investment in the oil sector. Mexico's oil production has declined over much of the past decade, from 3.3 MMbbl/d in 2005 to about 2.5 MMbbl/d in 2013, which was the lowest level since 1995. The reform is designed largely to take advantage of sizeable reserves in complex reservoirs that Pemex has neither the resources nor technology to develop effectively. The government recognizes the importance of stopping the declines in Mexican energy supply (an important source of government revenue) and expects the reforms to improve production over the next few decades.

To implement the reforms [37], the Mexican government has agreed to:

- Create four oil and gas exploration and production contract models, including service contracts, production-sharing, profit-sharing, and licenses
- Give Pemex first refusal on developing Mexican resources before private companies begin an initial bidding round (Round Zero), in which Pemex can provide financial and technical plans to develop the resources within three years
- Give regulatory authority over the oil and natural gas sectors to the Energy Regulatory Commission (CRE), the Secretaria de Energia de México (SENER), and the National Hydrocarbon Commission (CNH), and create a new National Agency of Industrial Safety and Environmental Protection
- Keep Pemex as state-owned but with more administrative and budgetary autonomy, and allow the company to compete with other firms for bids on new projects
- Establish the Mexican Petroleum Fund to manage contract payments and oil revenues.

The previous upstream contracting regime was one of entitlements *(asignaciones)*, in which a state-owned entity was granted the right to explore and produce hydrocarbons in a certain area for a certain period of time in exchange for tax payments. The four new contract models differ from the entitlements system and differ from each other in their fee and royalty structures. Service contracts are similar to the ones introduced as part of the previous (2008) effort to reform the Mexican energy sector. Under this arrangement, all crude produced is delivered by producers to the state in exchange for cash paid by the Mexican Petroleum Fund. Licenses, on the other hand, allow producers to take crude at the wellhead after making a series of payments to the state. The profit-sharing and production-sharing contracts, as well as licenses, will effectively allow producers to book reserves and reflect the potential value of the oil in their accounts—a particularly attractive incentive for investment in Mexico's energy sector [38]. It is expected that the different contract types will be applied according to the degree of risk associated with specific projects.

Mexico's Congress will need to specify the legislative details and implement the energy reforms in 2014, including the fiscal regime and contract terms for the various models, which will be crucial in determining whether foreign firms will invest. So-called *secondary legislation*—a package of 21 proposed legislative changes related to implementation of the reforms—was introduced in April 2014 [39] and approved in August 2014.

Pemex will remain state-owned but will be given more administrative and budgetary autonomy, and it will be permitted to compete with other firms for bids on new projects, enabling it to behave as a private firm rather than as a state-owned entity. Additionally, under the new contract regimes, Pemex can enter into joint ventures with private firms.

The Round Zero bidding process allowed Pemex to choose the projects it assesses to be technically and financially feasible within a three-year period. Round Zero began with the submission of Pemex's request to SENER in March 2014, proposing that the company retain 100% of existing producing areas, 83% of proven and probable reserves, and 31% of prospective resources [40]. SENER, in consultation with the CNH, assessed the feasibility of the Pemex request [41]. Round Zero results were published in August 2014, granting Pemex's requested portfolio of 100% of producing areas and 83% of proven and probable reserves, along with 21% of prospective resources [42].

The announcement of Round Zero results coincided with the start of the Round One bidding process. Preliminary terms will be released over the next several months, and the tender will conclude with a mid-2015 award of contracts. Foreign and local companies may bid on 109 exploration blocks and 60 extraction blocks [43].

The regulatory structure is also being updated to oversee the liberalized energy markets. The new structure should add checks and balances to the system, helping to keep it transparent and to discourage corruption. Exploration and production (E&P) contracts and entitlements will be regulated by the CNH and SENER, while midstream and downstream activities are under the jurisdiction of SENER and CRE. The Ministry of Finance is responsible for establishing economic and fiscal terms of the E&P contracts [44].

EIA takes a cautiously optimistic view of the potential for successful reform. In the IEO2014 Reference case, total liquid fuels output in Mexico remains steady, rather than sharply declining, over the next several years, stabilizing at 2.9 MMbbl/d through 2020 and

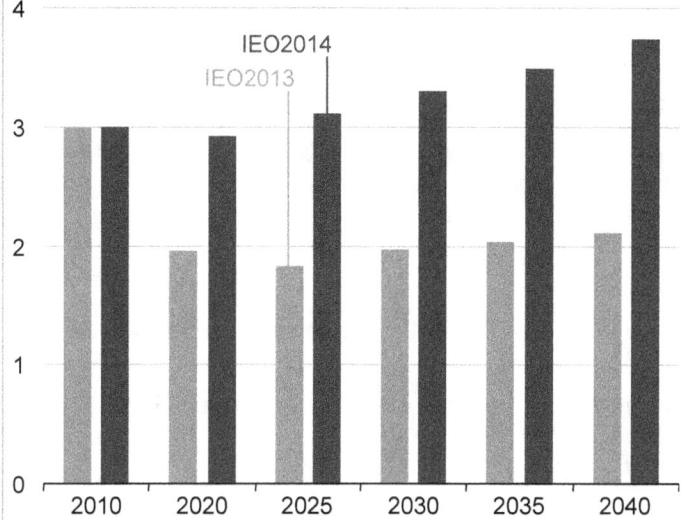

Figure 17. Mexico petroleum and other liquid fuels production in IEO2013 and IEO2014, 2010-40 (million barrels per day)

then rising to 3.7 MMbbl/d in 2040, or 1.7 MMbbl/d more than projected in the IEO2013 Reference case (Figure 17). There is, however, substantial upside uncertainty in the IEO2014 projections. With 10 billion barrels of proved oil reserves and potentially large volumes of hydrocarbon resources in Mexican territory in the deepwater Gulf of Mexico, the potential is enormous, and successful implementation of the energy reforms could substantially transform the outlook for Mexico's oil production.

Endnotes

Links current as of August 2014

1. U.S. Energy Information Administration, *Short-Term Energy Outlook June 2014* (June 10, 2014), www.eia.gov/forecasts/steo/archives/jun2014.pdf.
2. H. Luchnikava, "India," *IHS Economics: Monthly Outlook: Asia-Pacific* (April 2014), pp. 25-31, www.ihs.com (subscription site).
3. S. Bakir, IHS Connect, *Energy Country Profiles: Utilities: Electricity and Gas—Saudi Arabia* (August 1, 2013), www.ihs.com (subscription site).
4. S. Bakir and J. Ingram, "Iraq's deputy PM confirms reduced oil production targets," *IHS Energy: Oil & Gas Risk Service* (June 4, 2013), www.ihs.com/products/Global-Insight/industry-economic-report/aspx?ID=1065979820.
5. "Historical data: Oil production," *BP Statistical Review of World Energy* (June 2014), http://www.bp.com/en/global/corporate/about-bp/energy-economics/statistical-review-of-world-energy.html.
6. R. Alkadiri, "Split between US and EU likely if Congress is seen to play spoiler role in negotiations with Iran," *IHS Energy: Oil & Gas Risk Service* (March 14, 2014), www.ihs.com (subscription site).
7. "Interim Iran Deal Offers Some Respite for Oil Markets," *PFC Energy: Market Intelligence Service* (November 25, 2013), www.pfcenergy.com (subscription site).
8. U.S. Energy Information Administration, *Country Analysis Brief: Kuwait* (July 2013), http://www.eia.gov/countries/cab.cfm?fips=KU.
9. "Oil production begins at BP's PSVM project in offshore Angola," *Offshore Technology* (February 4, 2013), http://www.offshore-technology.com/news/newsoil-production-bp-psvm-project-offshore-angola/.
10. "Angola deepwater outlook and company positioning," *IHS Energy: Strategic Horizons Memo* (March 31, 2014), www.ihs.com (subscription site).
11. "Angola deepwater outlook and company positioning," *IHS Energy: Strategic Horizons Memo* (March 31, 2014), www.ihs.com (subscription site).
12. Libya: Ongoing political disarray to keep output low and volatile," *PFC Energy: National Oil Company Strategies Service Memo* (March 4, 2014), www.pfcenergy.com (subscription site).
13. C. Hunter, *Energy Country Profiles: Oil & Gas: Upstream—Algeria* (April 17, 2014), www.ihs.com (subscription site).
14. Venezuela: Pressures mount for new upstream approach," *IHS Energy: Strategic Horizons* (April 14, 2014), www.ihs.com (subscription site).
15. Government of Canada, National Energy Board, *ARCHIVED—Estimated Production of Canadian Crude Oil and Equivalent* (August 14, 2013), http://www.neb.gc.ca/clf-nsi/archives/rnrgynfmtn/sttstc/crdlndptrlmprdct/stmtdprdctnrchv-eng.html.
16. National Resources Canada, *North American Tight Light Oil* (April 7, 2014), http://www.nrcan.gc.ca/energy/crude-petroleum/4559.
17. J. Kerr, IHS Connect, *Energy Country Profiles: Oil & Gas: Upstream—Brazil* (November 21, 2013), www.ihs.com (subscription site).
18. J. Kerr, IHS Connect, *Energy Country Profiles: Oil & Gas: Upstream—Brazil* (November 21, 2013), www.ihs.com (subscription site).
19. FACTS Global Energy, *Latin America Oil Monthly* (May 22, 2014), pp. 14-15, www.fgenergy.com (subscription site).
20. J. Fargo, "Brazil sets minimum signing bonus of $6.6 billion for Libra," *Oil Daily* (July 8, 2013), www.energyintel.com (subscription site).
21. Deloitte, *First Brazilian Presalt Round 2013—Results* (October 21, 2013), http://www.psg.deloitte.com/newslicensingrounds_br_131114.asp.
22. U.S. Energy Information Administration, *Technically Recoverable Shale Oil and Shale Gas Resources: An Assessment of 137 Shale Formations in 41 Countries Outside the United States* (Washington, DC: June 2013).
23. J. Kerr, IHS Connect, *Energy Country Profiles: Oil & Gas: Upstream—Argentina* (May 16, 2014), www.ihs.com (subscription site).
24. FACTS Global Energy, *FSU Monthly Report* (May 16, 2014), pp. 1-2, www.fgenergy.com (subscription site).
25. "Kashagan production restart depends on pipelines investigation results," *Offshore Energy Today* (April 7, 2014), http://www.offshoreenergytoday.com/kashagan-production-restart-depends-on-pipelines-investigation-results/.
26. A. Neff, IHS Energy, *Energy Country Profiles: Oil & Gas: Upstream—Kazakhstan* (December 19, 2013), www.ihs.com (subscription site).
27. A. Neff, IHS Energy, *Energy Country Profiles: Oil & Gas: Upstream—Kazakhstan* (December 19, 2013), www.ihs.com (subscription site).

28. A. Neff, IHS Energy, *Energy Country Profiles: Oil & Gas: Upstream—Russia* (July 16, 2013), www.ihs.com (subscription site).
29. "Russia goes back to basics to boost output," *Petroleum Intelligence Weekly* (April 28, 2014), www.energyintel.com (subscription site).
30. A. Neff, "Tight oil tax breaks come into effect as Russia looks for future output growth from hard-to-recover oil," *IHS Energy: Oil & Gas Risk Service* (September 3, 2013), www.ihs.com (subscription site).
31. "Majors show up (and sign up) in St. Petersburg," *Petroleum Intelligence Weekly* (June 2, 2014), www.energyintel.com (subscription site).
32. S. Reed, "Energy companies rethinking Russia after new round of sanctions," *New York Times* (July 30, 2014), http://nyti.ms/1o5vdhQ.
33. C. Belghmidi, IHS Connect, *Energy Country Profiles: Oil & Gas: Upstream—United Kingdom* (July 23, 2013), www.ihs.com (subscription site).
34. "Energy Country Profiles—Norway," *IHS Energy: Oil & Gas Risk Service* (October 4, 2013), www.ihs.com (subscription site).
35. B. Koranyi, "Norway's rising oil costs hit Arctic output hopes," *Thomson Reuters* (January 16, 2014), www.uk.reuters.com; and "Johan Castberg (formerly Skrugard) Field Development Project, Barents Sea," *Offshore Technology Market & Customer Insight* (2014), www.offshore-technology.com/projects/skrugard-field-development-project-norway/.
36. S. Bakir, IHS Connect, *Energy Country Profiles: Oil & Gas: Upstream—India* (December 19, 2013), www.ihs.com (subscription site).
37. "Mexico's upstream opening: What is the potential?" *IHS Energy: Strategic Horizons* (February 26, 2014), www.ihs.com (subscription site).
38. M. Brown, "Analysis of proposed hydrocarbon legal regime in Mexico," *Legal Update* (May 30, 2014), http://www.mayerbrown.com/Analysis-of-Proposed-Hydrocarbon-Legal-Regime-in-Mexico-05-29-2014/.
39. N.B. Mendoza and J. Fargo, "Mexico presents laws to advance energy reform," *Oil Daily* (May 1, 2014), www.energyintel.com (subscription site).
40. "PEMEX nominates round zero areas in Mexican opening," *IHS Energy: Strategic Horizons* (April 2, 2014), www.ihs.com (subscription site).
41. P. Dittrick, "OTC: Pemex preparing for cultural transition as Mexico reforms energy laws," *Oil & Gas Journal* (May 6, 2014), www.ogj.com (subscription site).
42. "FACTBOX: Mexico's round zero and round one oil projects," *Reuters* (August 14, 2014), http://in.reuters.com/article/2014/08/13/mexico-reforms-energy-idINL2N0QJ2Z620140813.
43. A. Williams. "Pemex granted all probable reserves sought in oil opening," *Bloomberg* (August 13, 2014), http://www.bloomberg.com/news/2014-08-13/pemex-s-production-future-set-at-21-share-of-potential-deposits.html.
44. M. Brown. "Analysis of proposed hydrocarbon legal regime in Mexico," *Legal Update* (May 30, 2014), http://www.mayerbrown.com/Analysis-of-Proposed-Hydrocarbon-Legal-Regime-in-Mexico-05-29-2014/.

Data sources

Links current as of August 2014

Table 1. North Sea Brent crude oil spot prices in three cases, 2010-40: EIA, *Annual Energy Outlook 2014*, DOE/EIA-0383(2014) (Washington, DC, April 2014), www.eia.gov/aeo.

Figure 1. North Sea Brent crude oil spot prices in three cases, 1990-2040: EIA, *Annual Energy Outlook 2014*, DOE/EIA-0383(2014) (Washington, DC, April 2014), www.eia.gov/aeo.

Figure 2. World tight oil production in the Reference case, 2010 and 2040: History: EIA, Office of Energy Analysis, Office of Petroleum, Natural Gas & Biofuels Analysis. Projections: EIA, Generate World Oil Balance application (2014).

Figure 3. Liquid fuels consumption and production in three cases, 2040: Consumption: EIA, World Energy Projections System Plus (2014), runs 2014.03.21_100505 (Reference case), 2014.03.20_155716 (High Oil Price case), and 2014.03.24_145137 (Low Oil Price case). Production: EIA, Generate World Oil Balance application (2014) GWOB_IEO2014_RefCase.xlsx (Reference case), GWOB_IEO2014_HighPrice.xlsx (High Oil Price case), and GWOB_IEO2014_LowPrice.xlsx (Low Oil Price case).

Figure 4. Liquid fuels supply and demand and North Sea Brent crude oil equilibrium prices in three cases: Representation estimated by EIA, Office of Energy Markets and Financial Analysis, Macroeconomic Analysis Team.

Figure 5. OECD and Non-OECD petroleum and other liquid fuels consumption, Reference case, 1990-2040: History: EIA, International Energy Statistics database (as of November 2013), www.eia.gov/ies. Projections: EIA, World Energy Projection System Plus (2014), run 2014.03.21_100505.

Figure 6. Non-OECD petroleum and other liquid fuels consumption by region, Reference case, 1990-2040: History: EIA, International Energy Statistics database (as of November 2013), www.eia.gov/ies. Projections: EIA, World Energy Projection System Plus (2014), run 2014.03.21_100505.

Table 2. World gross domestic product by OECD and non-OECD in three oil price cases, 1990-2040: Oxford Economic Model (February 2014).

Table 3. World petroleum and other liquid fuels consumption by region, Reference case, 1980-2040: History: EIA, International Energy Statistics database (as of November 2013), www.eia.gov/ies. Projections: EIA, World Energy Projection System Plus (2014), run 2014.03.21_100505.

Figure 7. Petroleum and other liquid fuels consumption in China and the United States, Reference case, 1990-2040: History: EIA, International Energy Statistics database (as of November 2012), www.eia.gov/ies. Projections: United States: EIA, *Annual Energy Outlook 2014*, DOE/EIA-0383(2014) (Washington, DC, April 2014), www.eia.gov/aeo; China: EIA, World Energy Projection System Plus (2014), run 2014.03.21_100505.

Figure 8. Middle East liquid fuels consumption by end-use sector, 2010-40: 2010: Derived from EIA, International Energy Statistics database (as of November 2013), www.eia.gov/ies. Projections: EIA, World Energy Projection System Plus (2014), run 2014.03.21_100505.

Figure 9. Petroleum and other liquid fuels production by region and type in the Reference case, 2000-2040: History: EIA, Office of Energy Analysis, Office of Petroleum, Natural Gas & Biofuels Analysis. Projections: EIA, Generate World Oil Balance application (2014), run IEO2014_GWOB_RefCase.xlsx.

Figure 10. OPEC crude and lease condensate production by region in the Reference case, 2010 and 2040: 2010: EIA, Office of Energy Analysis, Office of Petroleum, Natural Gas & Biofuels Analysis. 2040: EIA, Generate World Oil Balance application (2014), run IEO2014_GWOB_RefCase.xlsx.

Figure 11. Non-OPEC crude and lease condensate production, 2010 and 2040: 2010: EIA, Office of Energy Analysis, Office of Petroleum, Natural Gas & Biofuels Analysis. 2040: EIA, Generate World Oil Balance application (2014), run IEO2014_GWOB_RefCase.xlsx.

Table 4. World liquid fuels production in the Reference case, 2010-40: 2010: EIA, Office of Energy Analysis, Office of Petroleum, Natural Gas & Biofuels Analysis. Projections: EIA, Generate World Oil Balance application (2014), run IEO2014_GWOB_RefCase.xlsx.

Figure 12. Brazil and Argentina crude and lease condensate production, 2010-40: 2010: EIA, Office of Energy Analysis, Office of Petroleum, Natural Gas & Biofuels Analysis. Projections: EIA, Generate World Oil Balance application (2014), run IEO2014_GWOB_RefCase.xlsx.

Figure 13. Non-OECD Europe and Eurasia crude and lease condensate production, 2010-40: 2010: EIA, Office of Energy Analysis, Office of Petroleum, Natural Gas & Biofuels Analysis. Projections: EIA, Generate World Oil Balance application (2014), run IEO2014_GWOB_RefCase.xlsx.

Figure 14. World other liquid fuels production by source, 2010 and 2040: 2010: EIA, Office of Energy Analysis, Office of Petroleum, Natural Gas & Biofuels Analysis. 2040: EIA, Generate World Oil Balance application (2014), run IEO2014_GWOB_RefCase.xlsx.

Table 5. World other liquid fuels production by fuel type, 2010-40: 2010: EIA, Office of Energy Analysis, Office of Petroleum, Natural Gas & Biofuels Analysis. **Projections:** EIA, Generate World Oil Balance application (2014), run IEO2014_GWOB_RefCase.xlsx.

Figure 15. World petroleum and other liquid fuels consumption by country grouping in three cases, 2010 and 2040: 2010: EIA, International Energy Statistics database (as of November 2013). **2040:** EIA, World Energy Projections System Plus (2014), runs 2014.03.21_100505 (Reference case), 2014.03.20_155716 (High Oil Price case), and 2014.03.24_145137 (Low Oil Price case).

Figure 16. World petroleum and other liquid fuels production in three cases, 2010 and 2040. 2010: EIA, Office of Energy Analysis, Office of Petroleum, Natural Gas & Biofuels Analysis. **2040:** EIA, Generate World Oil Balance application (2014) GWOB_IEO2014_RefCase.xlsx (Reference case), GWOB_IEO2014_HighPrice.xlsx (High Oil Price case), and GWOB_IEO2014_LowPrice.xlsx (Low Oil Price case).

Figure 17. Mexico petroleum and other liquid fuels production in IEO2013 and IEO2014, 2010-40: IEO2013: Derived from EIA, *International Energy Outlook 2013*, DOE/EIA-0484(2014) (Washington, DC, July 2013). **IEO2014:** EIA, Generate World Oil Balance application (2014), run GWOB_IEO2014_RefCase.xlsx.

THIS PAGE INTENTIONALLY LEFT BLANK

Appendix A
Reference case projections

This page intentionally left blank

Table A1. World gross domestic product (GDP) by region expressed in purchasing power parity, Reference case, 2009-40
(billion 2005 dollars)

Region	History		Projections					Average annual percent change, 2010-40
	2009	2010	2020	2025	2030	2035	2040	
OECD								
OECD Americas	**15,498**	**15,929**	**20,709**	**23,279**	**26,153**	**29,306**	**32,836**	**2.4**
United States[a]	12,758	13,063	16,753	18,769	21,139	23,751	26,670	2.4
Canada	1,165	1,202	1,533	1,716	1,902	2,098	2,303	2.2
Mexico/Chile	1,575	1,664	2,423	2,794	3,113	3,457	3,863	2.8
OECD Europe	**14,262**	**14,618**	**17,681**	**19,752**	**21,775**	**23,972**	**26,304**	**2.0**
OECD Asia	**5,791**	**6,062**	**7,320**	**7,934**	**8,431**	**8,865**	**9,216**	**1.4**
Japan	3,776	3,948	4,448	4,661	4,787	4,858	4,837	0.7
South Korea	1,244	1,323	1,838	2,110	2,345	2,560	2,771	2.5
Australia/NewZealand	771	790	1,034	1,163	1,298	1,448	1,608	2.4
Total OECD	35,551	36,609	45,711	50,965	56,358	62,143	68,357	2.1
Non-OECD								
Non-OECD Europe and Eurasia	**4,346**	**4,502**	**6,257**	**7,363**	**8,451**	**9,666**	**10,689**	**2.9**
Russia	1,938	2,022	2,741	3,168	3,545	3,954	4,192	2.5
Other	2,408	2,480	3,515	4,196	4,906	5,712	6,496	3.3
Non-OECD Asia	**16,628**	**18,206**	**33,695**	**44,746**	**57,688**	**71,655**	**85,314**	**5.3**
China	8,299	9,167	18,179	24,055	31,431	38,867	44,890	5.4
India	3,364	3,661	6,511	8,919	11,399	14,272	17,580	5.4
Other	4,965	5,379	9,006	11,772	14,858	18,516	22,845	4.9
Middle East	**2,263**	**2,292**	**3,498**	**4,292**	**5,162**	**6,144**	**7,213**	**3.9**
Africa	**3,780**	**3,963**	**6,188**	**7,830**	**9,977**	**12,720**	**16,148**	**4.8**
Central and South America	**4,623**	**4,927**	**6,794**	**8,024**	**9,394**	**10,971**	**12,759**	**3.2**
Brazil	1,833	1,971	2,600	3,069	3,615	4,267	5,015	3.2
Other	2,790	2,955	4,194	4,955	5,779	6,704	7,744	3.3
Total Non-OECD	31,640	33,889	56,432	72,255	90,672	111,157	132,123	4.6
Total World	67,192	70,498	102,142	123,220	147,030	173,300	200,479	3.5

[a]Includes the 50 States and the District of Columbia.
Note: Totals may not equal sum of components due to independent rounding.
Sources: Derived from Oxford Economic Model (February 2014), www.oxfordeconomics.com (subscription site); EIA, *Annual Energy Outlook 2014*, DOE/EIA-0383(2014) (Washington, DC: April 2014); and AEO2014 National Energy Modeling System, run REF2014.D102413A (Reference case), www.eia.gov/aeo.

Table A2. World liquids consumption by region, Reference case, 2009-40
(million barrels per day)

Region	History 2009	History 2010	Projections 2020	Projections 2025	Projections 2030	Projections 2035	Projections 2040	Average annual percent change, 2010-40
OECD								
OECD Americas	**23.1**	**23.5**	**24.3**	**24.0**	**23.6**	**23.4**	**23.5**	**0.0**
United States[a]	18.6	18.9	19.2	19.0	18.6	18.5	18.4	-0.1
Canada	2.2	2.2	2.3	2.2	2.2	2.2	2.1	-0.1
Mexico/Chile	2.4	2.4	2.7	2.8	2.8	2.8	2.9	0.7
OECD Europe	**15.0**	**14.8**	**14.1**	**14.1**	**14.0**	**13.9**	**14.0**	**-0.2**
OECD Asia	**7.7**	**7.7**	**8.0**	**7.9**	**7.7**	**7.4**	**7.2**	**-0.2**
Japan	4.4	4.4	4.3	4.2	4.0	3.9	3.6	-0.6
South Korea	2.2	2.3	2.6	2.6	2.5	2.5	2.4	0.2
Australia/NewZealand	1.1	1.1	1.2	1.1	1.1	1.1	1.1	0.1
Total OECD	**45.8**	**46.0**	**46.4**	**45.9**	**45.3**	**44.8**	**44.7**	**-0.1**
Non-OECD								
Non-OECD Europe and Eurasia	**4.8**	**4.8**	**5.5**	**5.5**	**5.6**	**5.7**	**5.6**	**0.5**
Russia	3.0	3.0	3.3	3.2	3.2	3.2	3.0	0.0
Other	1.8	1.8	2.2	2.3	2.4	2.5	2.6	1.2
Non-OECD Asia	**18.4**	**19.8**	**26.5**	**30.2**	**34.8**	**39.0**	**43.2**	**2.6**
China	8.5	9.3	13.1	14.7	16.9	18.8	20.0	2.6
India	3.1	3.3	4.3	4.9	5.5	6.1	6.8	2.5
Other	6.7	7.2	9.1	10.7	12.3	14.2	16.4	2.8
Middle East	**6.5**	**6.7**	**8.4**	**8.8**	**9.6**	**10.3**	**11.1**	**1.7**
Africa	**3.3**	**3.4**	**3.9**	**4.3**	**4.8**	**5.4**	**6.2**	**2.0**
Central and South America	**5.7**	**6.0**	**6.9**	**7.0**	**7.4**	**7.9**	**8.6**	**1.2**
Brazil	2.5	2.6	3.1	3.2	3.4	3.7	4.1	1.5
Other	3.3	3.4	3.8	3.8	4.0	4.2	4.5	0.9
Total Non-OECD	**38.7**	**40.7**	**51.2**	**55.9**	**62.1**	**68.3**	**74.7**	**2.0**
Total World	**84.5**	**86.8**	**97.6**	**101.8**	**107.4**	**113.1**	**119.4**	**1.1**

[a]Includes the 50 States and the District of Columbia.
Note: Totals may not equal sum of components due to independent rounding.
Sources: **History:** U.S. Energy Information Administration (EIA), International Energy Statistics database (as of November 2013), www.eia.gov/ies.
Projections: EIA, *Annual Energy Outlook 2014*, DOE/EIA-0383(2014) (Washington, DC: April 2014), AEO2014 National Energy Modeling System, run REF2014.D102413A, www.eia.gov/aeo; and World Energy Projection System Plus (2014), run 2014.03.21_100505 (Reference case).

Table A3. World petroleum and other liquids consumption by region and end-use sector, Reference case, 2010-40
(quadrillion Btu)

Region	History 2010	Projections 2020	Projections 2025	Projections 2030	Projections 2035	Projections 2040	Average annual percent change, 2010-40
OECD							
United States							
Residential	1.1	0.9	0.8	0.8	0.7	0.7	-1.8
Commercial	0.7	0.7	0.7	0.7	0.7	0.7	0.1
Industrial	8.1	9.6	9.9	10.1	10.1	10.1	0.7
Transportation	26.9	25.6	24.7	23.9	23.7	23.7	-0.4
Electricity	0.4	0.2	0.2	0.2	0.2	0.2	-2.5
Total	**37.2**	**36.9**	**36.3**	**35.7**	**35.4**	**35.4**	**-0.2**
Canada							
Residential	0.1	0.1	0.1	0.1	0.1	0.1	-0.2
Commercial	0.1	0.1	0.1	0.1	0.1	0.1	-0.1
Industrial	1.7	1.8	1.8	1.8	1.7	1.6	-0.1
Transportation	2.4	2.5	2.4	2.4	2.3	2.4	0.0
Electricity	0.0	0.0	0.0	0.0	0.0	0.0	-1.0
Total	**4.3**	**4.6**	**4.5**	**4.4**	**4.3**	**4.3**	**-0.1**
Mexico/Chile							
Residential	0.3	0.3	0.3	0.3	0.3	0.3	0.1
Commercial	0.1	0.1	0.1	0.1	0.1	0.1	0.2
Industrial	1.1	1.3	1.4	1.5	1.5	1.5	0.9
Transportation	2.7	3.2	3.3	3.3	3.3	3.5	0.9
Electricity	0.4	0.4	0.4	0.4	0.3	0.3	-1.0
Total	**4.7**	**5.3**	**5.5**	**5.5**	**5.5**	**5.7**	**0.7**
OECD Europe							
Residential	2.1	1.9	1.8	1.8	1.7	1.7	-0.7
Commercial	0.9	0.8	0.8	0.8	0.7	0.7	-0.8
Industrial	9.6	9.4	9.6	9.9	10.1	10.2	0.2
Transportation	17.7	16.8	16.6	16.3	16.1	16.3	-0.3
Electricity	0.4	0.4	0.4	0.3	0.3	0.3	-0.9
Total	**30.6**	**29.3**	**29.2**	**29.1**	**28.9**	**29.1**	**-0.2**
Japan							
Residential	0.6	0.5	0.5	0.5	0.4	0.4	-1.1
Commercial	0.7	0.6	0.6	0.6	0.6	0.5	-0.7
Industrial	3.6	3.9	4.0	3.9	3.8	3.6	0.0
Transportation	3.7	3.2	3.1	2.9	2.7	2.6	-1.2
Electricity	0.5	0.5	0.4	0.4	0.4	0.4	-1.0
Total	**9.0**	**8.7**	**8.5**	**8.2**	**7.9**	**7.5**	**-0.6**

See notes at end of table.

Table A3. World petroleum and other liquids consumption by region and end-use sector, Reference case, 2010-40 (continued) (quadrillion Btu)

Region	History	Projections					Average annual percent change, 2010-40
	2010	2020	2025	2030	2035	2040	
OECD (continued)							
South Korea							
Residential	0.1	0.1	0.1	0.1	0.1	0.1	-0.1
Commercial	0.1	0.1	0.1	0.1	0.1	0.1	-0.6
Industrial	2.5	2.8	2.7	2.6	2.5	2.3	-0.2
Transportation	1.8	2.1	2.2	2.3	2.3	2.3	1.0
Electricity	0.1	0.1	0.1	0.1	0.1	0.1	-1.0
Total	**4.6**	**5.3**	**5.3**	**5.2**	**5.1**	**5.0**	**0.3**
Australia/New Zealand							
Residential	0.0	0.0	0.0	0.0	0.0	0.0	-0.3
Commercial	0.0	0.0	0.0	0.0	0.0	0.0	-0.2
Industrial	0.6	0.6	0.6	0.6	0.6	0.6	-0.2
Transportation	1.6	1.7	1.6	1.6	1.6	1.7	0.3
Electricity	0.0	0.0	0.0	0.0	0.0	0.0	-0.9
Total	**2.2**	**2.3**	**2.3**	**2.3**	**2.3**	**2.3**	**0.2**
Total OECD							
Residential	4.3	3.8	3.7	3.6	3.4	3.3	-0.9
Commercial	2.6	2.5	2.4	2.4	2.3	2.3	-0.4
Industrial	27.2	29.4	30.0	30.3	30.2	29.8	0.3
Transportation	56.6	55.1	53.8	52.6	52.1	52.5	-0.3
Electricity	1.9	1.5	1.5	1.4	1.4	1.3	-1.2
Total	**92.6**	**92.4**	**91.4**	**90.3**	**89.4**	**89.3**	**-0.1**
Non-OECD							
Russia							
Residential	0.3	0.3	0.3	0.3	0.3	0.3	-0.7
Commercial	0.1	0.1	0.1	0.1	0.1	0.1	-1.5
Industrial	2.7	2.7	2.6	2.7	2.7	2.4	-0.3
Transportation	2.9	3.5	3.4	3.4	3.3	3.2	0.4
Electricity	0.1	0.1	0.1	0.1	0.1	0.1	-1.0
Total	**6.0**	**6.7**	**6.5**	**6.5**	**6.4**	**6.0**	**0.0**
Other Non-OECD Europe and Eurasia							
Residential	0.1	0.1	0.1	0.1	0.1	0.1	-0.1
Commercial	0.1	0.1	0.1	0.1	0.1	0.1	-0.4
Industrial	1.4	1.3	1.2	1.2	1.2	1.2	-0.6
Transportation	2.0	2.9	3.1	3.4	3.6	3.9	2.3
Electricity	0.1	0.1	0.1	0.1	0.1	0.1	-1.0
Total	**3.7**	**4.6**	**4.7**	**4.9**	**5.2**	**5.4**	**1.2**

See notes at end of table.

Table A3. World petroleum and other liquids consumption by region and end-use sector, Reference case, 2010-40 (continued) (quadrillion Btu)

Region	History 2010	Projections 2020	2025	2030	2035	2040	Average annual percent change, 2010-40
Non-OECD (continued)							
China							
Residential	1.2	1.2	1.1	1.0	1.0	0.9	-0.9
Commercial	1.1	1.0	1.0	1.0	0.9	0.9	-0.7
Industrial	8.4	9.5	10.0	11.1	11.9	12.2	1.3
Transportation	8.4	15.0	17.9	21.6	24.7	27.2	4.0
Electricity	0.1	0.1	0.0	0.0	0.0	0.0	-1.0
Total	**19.1**	**26.8**	**30.0**	**34.7**	**38.6**	**41.2**	**2.6**
India							
Residential	0.9	1.1	1.0	1.0	1.0	0.9	-0.1
Commercial	0.0	0.0	0.0	0.0	0.0	0.0	0.0
Industrial	3.2	3.6	4.0	4.6	5.2	5.7	2.0
Transportation	2.3	3.8	4.7	5.4	6.2	7.1	3.8
Electricity	0.2	0.2	0.2	0.2	0.2	0.2	-0.9
Total	**6.6**	**8.7**	**10.0**	**11.3**	**12.5**	**13.9**	**2.5**
Other Non-OECD Asia							
Residential	0.5	0.5	0.6	0.6	0.6	0.7	0.6
Commercial	0.3	0.3	0.3	0.4	0.4	0.4	1.1
Industrial	4.8	5.9	7.1	8.5	10.1	11.8	3.1
Transportation	8.2	11.1	13.2	15.2	17.5	20.4	3.1
Electricity	1.1	1.1	1.0	1.0	0.9	0.9	-0.8
Total	**14.9**	**18.9**	**22.2**	**25.7**	**29.5**	**34.2**	**2.8**
Middle East							
Residential	0.7	0.7	0.7	0.6	0.6	0.6	-0.5
Commercial	0.1	0.1	0.1	0.1	0.1	0.1	0.6
Industrial	3.8	5.2	5.8	7.2	8.6	10.1	3.3
Transportation	5.8	7.6	8.1	8.5	8.8	9.2	1.5
Electricity	3.4	3.4	3.3	3.1	3.0	2.8	-0.6
Total	**13.8**	**17.1**	**18.0**	**19.6**	**21.1**	**22.9**	**1.7**
Africa							
Residential	0.7	0.7	0.7	0.8	0.8	0.9	0.7
Commercial	0.1	0.1	0.1	0.1	0.1	0.1	1.5
Industrial	1.5	1.6	1.7	1.9	2.2	2.4	1.5
Transportation	3.7	4.8	5.4	6.2	7.2	8.5	2.8
Electricity	0.8	0.8	0.7	0.7	0.7	0.6	-0.8
Total	**6.9**	**8.0**	**8.7**	**9.7**	**11.0**	**12.6**	**2.0**

See notes at end of table.

Table A3. World petroleum and other liquids consumption by region and end-use sector, Reference case, 2010-40 (continued) (quadrillion Btu)

Region	History 2010	Projections 2020	Projections 2025	Projections 2030	Projections 2035	Projections 2040	Average annual percent change, 2010-40
Non-OECD (continued)							
Brazil							
Residential	0.3	0.3	0.3	0.3	0.3	0.3	0.1
Commercial	0.0	0.0	0.0	0.0	0.0	0.0	0.2
Industrial	2.0	2.3	2.2	2.4	2.8	3.2	1.4
Transportation	2.9	3.7	3.9	4.2	4.5	4.9	1.8
Electricity	0.1	0.1	0.1	0.1	0.1	0.1	-0.9
Total	5.4	6.4	6.6	7.1	7.7	8.5	1.5
Other Central and South America							
Residential	0.3	0.3	0.3	0.3	0.3	0.3	0.1
Commercial	0.1	0.1	0.1	0.1	0.1	0.1	0.7
Industrial	2.1	2.1	2.1	2.3	2.5	2.7	1.0
Transportation	3.4	4.2	4.4	4.7	4.9	5.2	1.4
Electricity	1.0	1.0	0.9	0.9	0.8	0.8	-0.9
Total	6.9	7.7	7.8	8.2	8.6	9.2	0.9
Total Non-OECD							
Residential	5.1	5.3	5.1	5.1	5.0	5.0	-0.1
Commercial	1.9	1.9	1.9	1.9	1.9	1.8	-0.1
Industrial	29.8	34.2	36.8	41.9	47.0	51.8	1.9
Transportation	39.6	56.6	64.3	72.6	80.7	89.6	2.8
Electricity	6.9	6.8	6.4	6.2	5.9	5.6	-0.7
Total	83.3	104.8	114.5	127.6	140.5	153.8	2.1
Total World							
Residential	9.5	9.2	8.8	8.6	8.5	8.3	-0.4
Commercial	4.5	4.4	4.3	4.2	4.2	4.1	-0.3
Industrial	57.0	63.7	66.8	72.2	77.3	81.6	1.2
Transportation	96.2	111.7	118.1	125.3	132.8	142.1	1.3
Electricity	8.8	8.3	7.9	7.6	7.2	6.9	-0.8
Total	176.0	197.2	206.0	217.9	229.9	243.1	1.1

Note: Totals may not equal sum of components due to independent rounding.
Sources: 2010: Derived from U.S. Energy Information Administration (EIA), International Energy Statistics database (as of November 2013), www.eia.gov/ies; and International Energy Agency, "Balances of OECD and Non-OECD Statistics" (2013), www.iea.org (subscription site).
Projections: EIA, *Annual Energy Outlook 2014*, DOE/EIA-0383(2014) (Washington, DC: April 2014), AEO2014 National Energy Modeling System, run REF2014.D102413A, www.eia.gov/aeo; and World Energy Projection System Plus (2014), run 2014.03.21_100505 (Reference case).

Table A4. World petroleum and other liquids production by region and country, Reference case, 2009-40
(million barrels per day)

Region	History			Projections					Average annual percent change, 2010-40
	2009	2010	2011	2020	2025	2030	2035	2040	
OPEC[a]	34.1	35.4	35.7	38.7	40.7	44.4	48.2	52.1	1.3
Middle East	23.2	24.3	25.9	27.1	28.8	32.2	35.5	38.8	1.6
North Africa	3.8	3.7	2.4	3.5	3.6	3.7	3.8	4.1	0.3
West Africa	4.1	4.5	4.4	5.0	5.2	5.2	5.3	5.4	0.7
South America	3.0	2.9	3.0	3.1	3.2	3.3	3.5	3.8	0.9
Non-OPEC	50.4	51.9	52.1	58.9	61.1	63.1	64.9	67.2	0.9
OECD	21.0	21.4	21.5	26.3	26.6	26.7	26.9	27.2	0.8
OECD North America	15.3	16.0	16.5	22.2	22.6	22.8	23.0	22.9	1.2
United States	8.9	9.4	9.8	14.2	13.9	13.2	12.9	12.4	0.9
Canada	3.4	3.6	3.7	5.0	5.7	6.2	6.6	6.7	2.1
Mexico and Chile	3.0	3.0	3.0	2.9	3.1	3.3	3.5	3.8	0.8
OECD Europe	4.9	4.6	4.3	3.3	3.0	2.9	2.9	3.1	-1.3
North Sea	3.9	3.6	3.3	2.2	1.9	1.8	1.7	1.9	-2.1
Other	1.0	1.0	1.0	1.0	1.1	1.1	1.1	1.2	0.6
OECD Asia	0.8	0.8	0.8	0.8	1.0	1.0	1.1	1.1	1.0
Australia and New Zealand	0.7	0.7	0.6	0.6	0.8	0.9	0.9	1.0	1.2
Other	0.2	0.2	0.2	0.2	0.2	0.2	0.2	0.2	0.1
Non-OECD	29.4	30.5	30.5	32.6	34.5	36.3	38.0	40.1	0.9
Non-OECD Europe and Eurasia	13.1	13.4	13.5	14.1	14.9	15.6	16.5	17.3	0.9
Russia	9.9	10.1	10.2	10.7	10.8	11.2	11.8	12.2	0.6
Caspian Area	2.8	3.0	3.0	3.2	4.0	4.3	4.5	4.9	1.7
Kazakhstan	1.5	1.6	1.6	2.1	2.9	3.1	3.2	3.5	2.5
Other	1.3	1.3	1.3	1.1	1.1	1.2	1.3	1.4	0.3
Other	0.3	0.3	0.3	0.2	0.2	0.2	0.2	0.3	-0.6
Non-OECD Asia	7.8	8.2	8.2	8.9	9.1	9.2	9.2	9.5	0.5
China	4.1	4.4	4.3	5.1	5.4	5.5	5.6	5.7	0.9
India	0.9	1.0	1.0	1.1	1.2	1.3	1.3	1.4	1.4
Other	2.9	2.9	2.9	2.7	2.5	2.3	2.3	2.4	-0.6
Middle East (Non-OPEC)	1.6	1.6	1.5	1.0	0.9	0.8	0.8	0.8	-2.4
Oman	0.8	0.9	0.9	0.8	0.7	0.6	0.5	0.5	-1.6
Other	0.7	0.7	0.6	0.3	0.3	0.3	0.2	0.2	-3.6
Africa	2.6	2.6	2.6	2.7	2.9	2.9	3.0	3.3	0.8
Ghana	0.0	0.0	0.1	0.2	0.2	0.2	0.2	0.2	11.6
Other	2.6	2.6	2.6	2.5	2.6	2.7	2.8	3.1	0.6
Central and South America	4.3	4.6	4.7	5.9	6.7	7.8	8.4	9.1	2.3
Brazil	2.4	2.5	2.5	3.2	4.0	4.8	5.3	5.6	2.7
Other	1.9	2.1	2.1	2.7	2.7	3.0	3.1	3.4	1.7
Total World	84.5	87.2	87.8	97.6	101.8	107.4	113.1	119.4	1.1
OPEC Share of World Production	40%	41%	41%	40%	40%	41%	43%	44%	
Persian Gulf Share of World Production	27%	28%	30%	28%	28%	30%	31%	33%	

[a]OPEC = Organization of the Petroleum Exporting Countries (OPEC-13).
Note: Totals may not equal sum of components due to independent rounding.
Sources: **History:** U.S. Energy Information Administration (EIA), Office of Energy Analysis and Office of Petroleum, Natural Gas & Biofuels Analysis. **Projections:** EIA, Generate World Oil Balance application (2014), run IEO2014_GWOB_RefCase.xlsx.

Appendix A

Table A5. World crude and lease condensate[a] production by region and country, Reference case, 2009-40
(million barrels per day)

Region	History			Projections					Average annual percent change, 2010-40
	2009	2010	2011	2020	2025	2030	2035	2040	
OPEC[b]	**31.0**	**32.0**	**32.2**	**34.4**	**36.1**	**39.5**	**42.9**	**46.2**	**1.2**
Middle East	20.8	21.7	23.0	23.8	25.2	28.4	31.5	34.5	1.6
North Africa	3.3	3.2	2.0	2.9	2.9	2.9	2.9	3.0	-0.3
West Africa	4.1	4.4	4.3	4.9	5.0	5.1	5.2	5.3	0.6
South America	2.8	2.7	2.8	2.9	2.9	3.0	3.2	3.5	0.9
Non-OPEC	**41.9**	**42.9**	**42.8**	**48.3**	**49.4**	**50.4**	**51.4**	**52.9**	**0.7**
OECD	**15.3**	**15.4**	**15.2**	**19.5**	**19.5**	**19.4**	**19.5**	**19.6**	**0.8**
OECD North America	10.8	11.2	11.5	16.8	17.0	17.1	17.2	17.2	1.4
United States	5.5	5.6	5.8	9.8	9.3	8.6	8.2	7.8	1.1
Canada	2.6	2.9	3.0	4.4	4.9	5.5	5.8	5.9	2.4
Mexico and Chile	2.7	2.6	2.6	2.6	2.8	3.0	3.2	3.5	0.9
OECD Europe	**3.9**	**3.6**	**3.3**	**2.2**	**1.8**	**1.7**	**1.6**	**1.7**	**-2.5**
North Sea	3.4	3.1	2.8	1.8	1.5	1.3	1.3	1.4	-2.6
Other	0.5	0.5	0.5	0.4	0.3	0.3	0.3	0.3	-2.2
OECD Asia	**0.5**	**0.5**	**0.5**	**0.5**	**0.6**	**0.7**	**0.7**	**0.8**	**1.1**
Australia and New Zealand	0.5	0.5	0.5	0.5	0.6	0.7	0.7	0.8	1.1
Other	0.0	0.0	0.0	0.0	0.0	0.0	0.0	0.0	-1.7
Non-OECD	**26.6**	**27.5**	**27.5**	**28.8**	**29.9**	**31.0**	**31.9**	**33.2**	**0.6**
Non-OECD Europe and Eurasia	12.4	12.7	12.8	13.3	13.9	14.4	15.1	15.8	0.7
Russia	9.5	9.7	9.8	10.2	10.1	10.4	10.8	11.1	0.5
Caspian Area	**2.7**	**2.8**	**2.8**	**3.0**	**3.7**	**3.9**	**4.1**	**4.5**	**1.6**
Kazakhstan	1.5	1.6	1.6	1.9	2.7	2.8	2.9	3.1	2.3
Other	1.2	1.3	1.2	1.1	1.0	1.1	1.2	1.4	0.3
Other	**0.3**	**0.2**	**0.2**	**0.2**	**0.1**	**0.1**	**0.1**	**0.2**	**-0.9**
Non-OECD Asia	**6.9**	**7.3**	**7.2**	**7.5**	**7.3**	**7.0**	**6.7**	**6.6**	**-0.3**
China	3.8	4.1	4.1	4.5	4.7	4.6	4.4	4.1	0.0
India	0.7	0.7	0.8	0.8	0.7	0.7	0.8	0.8	0.3
Other	2.4	2.4	2.3	2.2	1.9	1.7	1.6	1.7	-1.3
Middle East (Non-OPEC)	**1.5**	**1.5**	**1.5**	**1.0**	**0.9**	**0.8**	**0.8**	**0.7**	**-2.4**
Africa	**2.2**	**2.2**	**2.2**	**2.2**	**2.3**	**2.4**	**2.5**	**2.7**	**0.7**
Central and South America	**3.6**	**3.8**	**3.9**	**4.8**	**5.5**	**6.3**	**6.9**	**7.4**	**2.3**
Brazil	2.0	2.1	2.1	2.6	3.2	3.8	4.2	4.5	2.6
Other	1.6	1.7	1.8	2.2	2.3	2.5	2.7	2.9	1.8
Total World	**72.9**	**74.9**	**75.0**	**82.7**	**85.5**	**89.9**	**94.3**	**99.1**	**0.9**
OPEC Share of World Production	42%	43%	43%	42%	42%	44%	45%	47%	
Persian Gulf Share of World Production	29%	29%	31%	29%	30%	32%	33%	35%	

[a] Crude and lease condensate includes tight oil, shale oil, extra-heavy oil, field condensate, and bitumen.
[b] OPEC = Organization of the Petroleum Exporting Countries (OPEC-13).
Note: Totals may not equal sum of components due to independent rounding.
Sources: **History:** U.S. Energy Information Administration (EIA), Office of Energy Analysis and Office of Petroleum, Natural Gas & Biofuels Analysis. **Projections:** EIA, Generate World Oil Balance application (2014), run IEO2014_GWOB_RefCase.xlsx.

Table A6. World other liquid fuels[a] production by region and country, Reference case, 2009-40
(million barrels per day)

Region	History			Projections					Average annual percent change, 2010-40
	2009	2010	2011	2020	2025	2030	2035	2040	
OPEC[b]	**3.1**	**3.3**	**3.5**	**4.3**	**4.6**	**4.9**	**5.3**	**5.9**	**1.9**
Natural gas plant liquids	3.1	3.3	3.4	4.0	4.2	4.5	4.9	5.4	1.7
Biofuels[c]	0.0	0.0	0.0	0.0	0.0	0.0	0.0	0.0	—
Coal-to-liquids	0.0	0.0	0.0	0.0	0.0	0.0	0.0	0.0	—
Gas-to-liquids (primarily Qatar)	0.0	0.0	0.1	0.3	0.3	0.4	0.4	0.4	14.1
Refinery gain	0.0	0.0	0.0	0.0	0.0	0.0	0.1	0.1	0.9
Non-OPEC	**8.5**	**9.0**	**9.3**	**10.6**	**11.7**	**12.6**	**13.5**	**14.4**	**1.6**
OECD	5.8	6.1	6.3	6.8	7.1	7.3	7.4	7.6	0.7
Natural gas plant liquids	3.4	3.5	3.6	4.0	4.2	4.3	4.3	4.4	0.8
Biofuels[c]	0.8	0.8	1.0	1.1	1.2	1.2	1.3	1.3	1.6
Coal-to-liquids	0.0	0.0	0.0	0.0	0.0	0.0	0.0	0.0	12.7
Gas-to-liquids	0.0	0.0	0.0	0.0	0.1	0.1	0.1	0.1	—
Kerogen	0.0	0.0	0.0	0.0	0.0	0.0	0.0	0.0	0.6
Refinery gain	1.6	1.7	1.7	1.7	1.7	1.7	1.7	1.7	0.0
Non-OECD	2.7	2.9	3.0	3.8	4.6	5.4	6.1	6.8	2.8
Natural gas plant liquids	1.6	1.6	1.7	1.9	2.2	2.4	2.7	2.9	1.9
Biofuels[c]	0.4	0.5	0.5	0.7	0.9	1.2	1.4	1.6	4.1
Coal-to-liquids	0.2	0.2	0.2	0.3	0.5	0.7	0.8	1.0	6.1
Gas-to-liquids	0.1	0.1	0.1	0.1	0.1	0.1	0.1	0.1	2.0
Refinery gain	0.5	0.6	0.6	0.8	0.9	1.0	1.1	1.2	2.3
Total World	**11.6**	**12.3**	**12.8**	**14.9**	**16.3**	**17.6**	**18.8**	**20.3**	**1.7**
Natural Gas Plant Liquids	**8.1**	**8.4**	**8.7**	**9.9**	**10.6**	**11.2**	**11.9**	**12.7**	**1.4**
United States	1.9	2.1	2.2	2.6	2.9	3.0	3.0	3.0	1.2
Russia	0.4	0.4	0.4	0.5	0.6	0.8	0.9	1.0	2.9
Biofuels[c]	**1.2**	**1.3**	**1.5**	**1.8**	**2.1**	**2.4**	**2.7**	**3.0**	**2.7**
Brazil	0.3	0.3	0.3	0.4	0.5	0.7	0.8	0.8	3.0
China	0.0	0.0	0.0	0.1	0.1	0.2	0.3	0.4	9.2
India	0.0	0.0	0.0	0.0	0.0	0.0	0.0	0.0	7.8
United States	0.5	0.6	0.7	0.7	0.7	0.7	0.7	0.7	0.5
Coal-to-liquids	**0.2**	**0.2**	**0.2**	**0.3**	**0.5**	**0.7**	**0.9**	**1.1**	**6.2**
Australia/New Zealand	0.0	0.0	0.0	0.0	0.0	0.0	0.0	0.0	—
China	0.0	0.0	0.0	0.1	0.2	0.3	0.5	0.6	14.9
Germany	0.0	0.0	0.0	0.0	0.0	0.0	0.0	0.0	0.3
India	0.0	0.0	0.0	0.0	0.1	0.1	0.1	0.2	—
South Africa	0.2	0.2	0.2	0.2	0.2	0.2	0.2	0.2	1.0
United States	0.0	0.0	0.0	0.0	0.0	0.0	0.0	0.0	—
Gas-to-liquids	**0.1**	**0.1**	**0.1**	**0.3**	**0.4**	**0.5**	**0.6**	**0.6**	**7.6**
Qatar	0.0	0.0	0.1	0.2	0.3	0.3	0.4	0.4	13.6
South Africa	0.0	0.0	0.0	0.0	0.0	0.0	0.1	0.1	0.7
Refinery Gain	**2.2**	**2.3**	**2.4**	**2.5**	**2.6**	**2.7**	**2.8**	**2.9**	**0.8**
United States	1.0	1.1	1.1	1.1	1.0	1.0	0.9	1.0	-0.4
China	0.2	0.2	0.2	0.3	0.4	0.4	0.5	0.5	2.6

[a]Other liquid fuels include natural gas plant liquids, biofuels, gas-to-liquids, coal-to-liquids, kerogen, and refinery gain.
[b]OPEC = Organization of the Petroleum Exporting Countries (OPEC-13).
[c]Ethanol values are reported on a gasoline-equivalent basis.
Note: Totals may not equal sum of components due to independent rounding.
Sources: **History:** U.S. Energy Information Administration (EIA), Office of Energy Analysis and Office of Petroleum, Natural Gas & Biofuels Analysis. **Projections:** EIA, Generate World Oil Balance application (2014), run IEO2014_GWOB_RefCase.xlsx.

THIS PAGE INTENTIONALLY LEFT BLANK

Appendix B
High Oil Price case projections

THIS PAGE INTENTIONALLY LEFT BLANK

Table B1. World gross domestic product (GDP) by region expressed in purchasing power parity, High Oil Price case, 2009-40
(billion 2005 dollars)

Region	History		Projections					Average annual percent change, 2010-40
	2009	2010	2020	2025	2030	2035	2040	
OECD								
OECD Americas	**15,498**	**15,929**	**20,779**	**23,465**	**26,267**	**29,532**	**33,262**	**2.5**
United States[a]	12,758	13,063	16,812	18,894	21,100	23,753	26,772	2.4
Canada	1,165	1,202	1,532	1,722	1,917	2,123	2,342	2.2
Mexico/Chile	1,575	1,664	2,435	2,849	3,250	3,656	4,149	3.1
OECD Europe	**14,262**	**14,618**	**17,602**	**19,738**	**21,731**	**23,983**	**26,351**	**2.0**
OECD Asia	**5,791**	**6,062**	**7,263**	**7,923**	**8,435**	**8,916**	**9,337**	**1.5**
Japan	3,776	3,948	4,398	4,641	4,770	4,846	4,830	0.7
South Korea	1,244	1,323	1,830	2,116	2,362	2,612	2,883	2.6
Australia/NewZealand	771	790	1,035	1,165	1,303	1,458	1,625	2.4
Total OECD	**35,551**	**36,609**	**45,643**	**51,126**	**56,434**	**62,431**	**68,950**	**2.1**
Non-OECD								
Non-OECD Europe and Eurasia	**4,346**	**4,502**	**6,360**	**7,445**	**8,546**	**9,710**	**10,781**	**3.0**
Russia	1,938	2,022	2,812	3,218	3,605	3,976	4,228	2.5
Other	2,408	2,480	3,547	4,227	4,941	5,734	6,553	3.3
Non-OECD Asia	**16,628**	**18,206**	**34,225**	**46,407**	**61,648**	**79,641**	**98,380**	**5.8**
China	8,299	9,167	18,764	25,689	35,194	46,113	56,190	6.2
India	3,364	3,661	6,505	8,912	11,428	14,522	18,279	5.5
Other	4,965	5,379	8,955	11,806	15,026	19,006	23,911	5.1
Middle East	**2,263**	**2,292**	**3,539**	**4,386**	**5,338**	**6,401**	**7,622**	**4.1**
Africa	**3,780**	**3,963**	**6,136**	**7,856**	**10,107**	**13,026**	**16,728**	**4.9**
Central and South America	**4,623**	**4,927**	**6,692**	**7,904**	**9,311**	**10,989**	**12,921**	**3.3**
Brazil	1,833	1,971	2,545	3,001	3,565	4,262	5,076	3.2
Other	2,790	2,955	4,147	4,904	5,746	6,727	7,845	3.3
Total Non-OECD	**31,640**	**33,889**	**56,952**	**73,998**	**94,949**	**119,767**	**146,431**	**5.0**
Total World	**67,192**	**70,498**	**102,595**	**125,124**	**151,383**	**182,197**	**215,381**	**3.8**

[a]Includes the 50 States and the District of Columbia.
Note: Totals may not equal sum of components due to independent rounding.
Sources: Derived from Oxford Economic Model (February 2014), www.oxfordeconomics.com (subscription site); EIA, *Annual Energy Outlook 2014*, DOE/EIA-0383(2014) (Washington, DC: April 2014); and AEO2014 National Energy Modeling System, run HIGHPRICE.D120613A (High Oil Price case), www.eia.gov/aeo.

Table B2. World liquids consumption by region, High Oil Price case, 2009-40
(million barrels per day)

Region	History 2009	History 2010	Projections 2020	Projections 2025	Projections 2030	Projections 2035	Projections 2040	Average annual percent change, 2010-40
OECD								
OECD Americas	**23.1**	**23.5**	**23.4**	**22.9**	**22.3**	**22.0**	**22.0**	**-0.2**
United States[a]	18.6	18.9	18.6	18.0	17.4	17.2	17.2	-0.3
Canada	2.2	2.2	2.2	2.1	2.0	2.0	1.9	-0.5
Mexico/Chile	2.4	2.4	2.6	2.7	2.8	2.8	2.9	0.6
OECD Europe	**15.0**	**14.8**	**13.7**	**13.7**	**13.6**	**13.5**	**13.4**	**-0.3**
OECD Asia	**7.7**	**7.7**	**7.6**	**7.5**	**7.2**	**6.9**	**6.7**	**-0.5**
Japan	4.4	4.4	4.0	4.0	3.8	3.6	3.3	-0.9
South Korea	2.2	2.3	2.4	2.4	2.4	2.3	2.2	0.0
Australia/NewZealand	1.1	1.1	1.1	1.1	1.1	1.1	1.1	0.0
Total OECD	**45.8**	**46.0**	**44.6**	**44.1**	**43.1**	**42.4**	**42.0**	**-0.3**
Non-OECD								
Non-OECD Europe and Eurasia	**4.8**	**4.8**	**5.4**	**5.3**	**5.3**	**5.3**	**5.1**	**0.2**
Russia	3.0	3.0	3.2	3.1	3.0	2.9	2.7	-0.4
Other	1.8	1.8	2.2	2.2	2.3	2.4	2.5	1.0
Non-OECD Asia	**18.4**	**19.8**	**26.0**	**30.4**	**35.8**	**41.6**	**47.4**	**3.0**
China	8.5	9.3	13.0	15.1	18.2	21.3	23.8	3.2
India	3.1	3.3	4.1	4.7	5.4	6.1	6.9	2.5
Other	6.7	7.2	8.8	10.5	12.2	14.2	16.7	2.9
Middle East	**6.5**	**6.7**	**8.4**	**9.1**	**10.2**	**11.4**	**12.7**	**2.1**
Africa	**3.3**	**3.4**	**3.8**	**4.2**	**4.7**	**5.4**	**6.3**	**2.1**
Central and South America	**5.7**	**6.0**	**6.6**	**6.8**	**7.2**	**7.8**	**8.5**	**1.1**
Brazil	2.5	2.6	3.0	3.0	3.3	3.6	4.1	1.5
Other	3.3	3.4	3.6	3.7	3.9	4.1	4.4	0.9
Total Non-OECD	**38.7**	**40.7**	**50.2**	**55.8**	**63.3**	**71.4**	**80.1**	**2.3**
Total World	**84.5**	**86.8**	**94.8**	**99.8**	**106.4**	**113.8**	**122.1**	**1.1**

[a]Includes the 50 States and the District of Columbia.
Note: Totals may not equal sum of components due to independent rounding.
Sources: **History:** U.S. Energy Information Administration (EIA), International Energy Statistics database (as of November 2013), www.eia.gov/ies.
Projections: EIA, *Annual Energy Outlook 2014*, DOE/EIA-0383(2014) (Washington, DC: April 2014), AEO2014 National Energy Modeling System, run HIGHPRICE.D120613A, www.eia.gov/aeo; and World Energy Projection System Plus (2014), run 2014.03.30_155716 (High Oil Price case).

Table B3. World petroleum and other liquids consumption by region and end-use sector, High Oil Price case, 2010-40 (quadrillion Btu)

Region	History 2010	Projections 2020	2025	2030	2035	2040	Average annual percent change, 2010-40
OECD							
United States							
Residential	1.1	0.9	0.8	0.7	0.7	0.6	-1.9
Commercial	0.7	0.6	0.6	0.6	0.6	0.6	-0.3
Industrial	8.1	9.4	9.9	9.9	10.0	10.0	0.7
Transportation	26.9	24.4	22.9	21.7	21.2	21.2	-0.8
Electricity	0.4	0.2	0.2	0.2	0.2	0.2	-2.5
Total	**37.2**	**35.4**	**34.3**	**33.1**	**32.7**	**32.7**	**-0.4**
Canada							
Residential	0.1	0.1	0.1	0.1	0.1	0.1	-0.4
Commercial	0.1	0.1	0.1	0.1	0.1	0.1	-0.3
Industrial	1.7	1.6	1.6	1.5	1.4	1.2	-1.1
Transportation	2.4	2.5	2.4	2.4	2.4	2.4	0.0
Electricity	0.0	0.0	0.0	0.0	0.0	0.0	-1.0
Total	**4.3**	**4.4**	**4.2**	**4.1**	**4.0**	**3.9**	**-0.4**
Mexico/Chile							
Residential	0.3	0.3	0.3	0.3	0.3	0.3	-0.1
Commercial	0.1	0.1	0.1	0.1	0.1	0.1	0.0
Industrial	1.1	1.2	1.3	1.3	1.3	1.2	0.1
Transportation	2.7	3.2	3.3	3.4	3.5	3.8	1.1
Electricity	0.4	0.4	0.4	0.4	0.3	0.3	-1.0
Total	**4.7**	**5.2**	**5.4**	**5.5**	**5.5**	**5.7**	**0.6**
OECD Europe							
Residential	2.1	1.8	1.7	1.7	1.6	1.6	-0.9
Commercial	0.9	0.8	0.7	0.7	0.7	0.7	-1.1
Industrial	9.6	8.9	9.2	9.3	9.4	9.1	-0.2
Transportation	17.7	16.5	16.5	16.2	16.0	16.2	-0.3
Electricity	0.4	0.4	0.4	0.3	0.3	0.3	-0.9
Total	**30.6**	**28.3**	**28.5**	**28.2**	**28.0**	**27.9**	**-0.3**
Japan							
Residential	0.6	0.5	0.5	0.4	0.4	0.4	-1.3
Commercial	0.7	0.6	0.5	0.5	0.5	0.5	-1.0
Industrial	3.6	3.6	3.7	3.5	3.3	3.0	-0.6
Transportation	3.7	3.1	3.0	2.9	2.7	2.6	-1.2
Electricity	0.5	0.5	0.4	0.4	0.4	0.4	-1.0
Total	**9.0**	**8.2**	**8.1**	**7.7**	**7.3**	**6.8**	**-0.9**

See notes at end of table.

Table B3. World petroleum and other liquids consumption by region and end-use sector, High Oil Price case, 2010-40 (continued) (quadrillion Btu)

Region	History	Projections					Average annual percent change, 2010-40
	2010	2020	2025	2030	2035	2040	
OECD (continued)							
South Korea							
Residential	0.1	0.1	0.1	0.1	0.1	0.1	-0.4
Commercial	0.1	0.1	0.1	0.1	0.1	0.1	-0.9
Industrial	2.5	2.5	2.4	2.3	2.1	1.9	-0.8
Transportation	1.8	2.1	2.2	2.2	2.3	2.4	1.0
Electricity	0.1	0.1	0.1	0.1	0.1	0.1	-1.0
Total	**4.6**	**5.0**	**5.0**	**4.9**	**4.7**	**4.6**	**0.0**
Australia/New Zealand							
Residential	0.0	0.0	0.0	0.0	0.0	0.0	-0.6
Commercial	0.0	0.0	0.0	0.0	0.0	0.0	-0.5
Industrial	0.6	0.6	0.5	0.5	0.5	0.5	-0.6
Transportation	1.6	1.6	1.6	1.6	1.6	1.7	0.3
Electricity	0.0	0.0	0.0	0.0	0.0	0.0	-0.9
Total	**2.2**	**2.3**	**2.2**	**2.2**	**2.2**	**2.3**	**0.1**
Total OECD							
Residential	4.3	3.6	3.5	3.3	3.2	3.1	-1.1
Commercial	2.6	2.2	2.2	2.2	2.1	2.1	-0.7
Industrial	27.2	27.9	28.6	28.5	28.0	26.9	0.0
Transportation	56.6	53.4	51.9	50.4	49.7	50.3	-0.4
Electricity	1.9	1.5	1.5	1.4	1.4	1.3	-1.2
Total	**92.6**	**88.8**	**87.6**	**85.8**	**84.3**	**83.8**	**-0.3**
Non-OECD							
Russia							
Residential	0.3	0.3	0.3	0.3	0.3	0.2	-0.9
Commercial	0.1	0.1	0.1	0.1	0.1	0.1	-1.8
Industrial	2.7	2.5	2.3	2.3	2.1	1.8	-1.3
Transportation	2.9	3.5	3.5	3.4	3.3	3.2	0.4
Electricity	0.1	0.1	0.1	0.1	0.1	0.1	-1.0
Total	**6.0**	**6.4**	**6.2**	**6.0**	**5.8**	**5.4**	**-0.4**
Other Non-OECD Europe and Eurasia							
Residential	0.1	0.1	0.1	0.1	0.1	0.1	-0.3
Commercial	0.1	0.1	0.1	0.1	0.1	0.1	-0.7
Industrial	1.4	1.3	1.2	1.1	1.0	0.9	-1.4
Transportation	2.0	2.9	3.1	3.4	3.6	3.9	2.3
Electricity	0.1	0.1	0.1	0.1	0.1	0.1	-1.0
Total	**3.7**	**4.5**	**4.6**	**4.8**	**5.0**	**5.1**	**1.1**

See notes at end of table.

Table B3. World petroleum and other liquids consumption by region and end-use sector, High Oil Price case, 2010-40 (continued) (quadrillion Btu)

Region	History 2010	Projections 2020	2025	2030	2035	2040	Average annual percent change, 2010-40
Non-OECD (continued)							
China							
Residential	1.2	1.1	1.0	1.0	1.0	0.9	-1.0
Commercial	1.1	1.0	0.9	0.9	0.9	0.9	-0.7
Industrial	8.4	9.2	9.9	11.3	12.5	13.0	1.5
Transportation	8.4	15.2	19.0	24.1	29.4	34.3	4.8
Electricity	0.1	0.1	0.0	0.0	0.0	0.0	-1.0
Total	**19.1**	**26.6**	**30.9**	**37.3**	**43.8**	**49.0**	**3.2**
India							
Residential	0.9	1.0	0.9	0.9	0.9	0.9	-0.4
Commercial	0.0	0.0	0.0	0.0	0.0	0.0	0.0
Industrial	3.2	3.4	3.8	4.4	5.1	5.8	2.0
Transportation	2.3	3.7	4.7	5.4	6.2	7.3	3.9
Electricity	0.2	0.2	0.2	0.2	0.2	0.2	-0.9
Total	**6.6**	**8.4**	**9.6**	**11.0**	**12.4**	**14.1**	**2.5**
Other Non-OECD Asia							
Residential	0.5	0.5	0.5	0.6	0.6	0.6	0.4
Commercial	0.3	0.3	0.3	0.3	0.4	0.4	0.9
Industrial	4.8	5.7	7.0	8.3	10.0	11.8	3.1
Transportation	8.2	10.8	13.1	15.3	17.8	21.2	3.2
Electricity	1.1	1.1	1.0	1.0	0.9	0.9	-0.8
Total	**14.9**	**18.4**	**21.9**	**25.5**	**29.6**	**34.9**	**2.9**
Middle East							
Residential	0.7	0.7	0.6	0.6	0.6	0.6	-0.7
Commercial	0.1	0.1	0.1	0.1	0.1	0.1	0.4
Industrial	3.8	5.3	6.2	8.0	10.1	12.3	4.0
Transportation	5.8	7.7	8.3	8.8	9.2	9.7	1.7
Electricity	3.4	3.4	3.3	3.1	3.0	2.8	-0.6
Total	**13.8**	**17.2**	**18.5**	**20.7**	**22.9**	**25.5**	**2.1**
Africa							
Residential	0.7	0.7	0.7	0.7	0.8	0.8	0.5
Commercial	0.1	0.1	0.1	0.1	0.1	0.1	1.3
Industrial	1.5	1.6	1.7	1.9	2.1	2.4	1.5
Transportation	3.7	4.7	5.4	6.3	7.3	8.8	2.9
Electricity	0.8	0.8	0.7	0.7	0.7	0.6	-0.8
Total	**6.9**	**7.8**	**8.6**	**9.7**	**11.0**	**12.8**	**2.1**

See notes at end of table.

Table B3. World petroleum and other liquids consumption by region and end-use sector, High Oil Price case, 2010-40 (continued) (quadrillion Btu)

Region	History 2010	Projections 2020	Projections 2025	Projections 2030	Projections 2035	Projections 2040	Average annual percent change, 2010-40
Non-OECD (continued)							
Brazil							
Residential	0.3	0.3	0.3	0.3	0.3	0.3	-0.1
Commercial	0.0	0.0	0.0	0.0	0.0	0.0	-0.1
Industrial	2.0	2.1	2.0	2.2	2.6	3.0	1.3
Transportation	2.9	3.6	3.8	4.1	4.5	4.9	1.8
Electricity	0.1	0.1	0.1	0.1	0.1	0.1	-0.9
Total	5.4	6.1	6.2	6.8	7.5	8.4	1.5
Other Central and South America							
Residential	0.3	0.3	0.3	0.3	0.3	0.3	-0.2
Commercial	0.1	0.1	0.1	0.1	0.1	0.1	0.4
Industrial	2.1	2.0	2.0	2.2	2.4	2.7	0.9
Transportation	3.4	4.1	4.4	4.6	4.9	5.2	1.4
Electricity	1.0	1.0	0.9	0.9	0.8	0.8	-0.9
Total	6.9	7.5	7.6	8.0	8.5	9.1	0.9
Total Non-OECD							
Residential	5.1	4.9	4.7	4.7	4.7	4.7	-0.3
Commercial	1.9	1.8	1.7	1.8	1.8	1.8	-0.2
Industrial	29.8	33.1	36.1	41.8	47.9	53.6	2.0
Transportation	39.6	56.2	65.3	75.3	86.2	98.6	3.1
Electricity	6.9	6.8	6.4	6.2	5.9	5.6	-0.7
Total	83.3	102.8	114.3	129.7	146.4	164.2	2.3
Total World							
Residential	9.5	8.5	8.2	8.1	8.0	7.8	-0.6
Commercial	4.5	4.0	3.9	3.9	3.9	3.8	-0.5
Industrial	57.0	61.0	64.7	70.3	75.8	80.6	1.2
Transportation	96.2	109.7	117.2	125.6	135.8	148.8	1.5
Electricity	8.8	8.3	7.9	7.6	7.2	6.9	-0.8
Total	176.0	191.5	201.9	215.5	230.7	248.0	1.2

Note: Totals may not equal sum of components due to independent rounding.
Sources: 2010: Derived from U.S. Energy Information Administration (EIA), International Energy Statistics database (as of November 2013), www.eia.gov/ies; and International Energy Agency, "Balances of OECD and Non-OECD Statistics" (2013), www.iea.org (subscription site).
Projections: EIA, *Annual Energy Outlook 2014*, DOE/EIA-0383(2014) (Washington, DC: April 2014), AEO2014 National Energy Modeling System, run HIGHPRICE.D120613A, www.eia.gov/aeo; and World Energy Projection System Plus (2014), run 2014.03.20_155716 (High Oil Price case).

Table B4. World petroleum and other liquids production by region and country, High Oil Price case, 2009-40
(million barrels per day)

Region	History			Projections					Average annual percent change, 2010-40
	2009	2010	2011	2020	2025	2030	2035	2040	
OPEC[a]	**34.1**	**35.4**	**35.7**	**33.1**	**34.5**	**37.8**	**41.0**	**43.7**	**0.7**
Middle East	**23.2**	**24.3**	**25.9**	**22.6**	**23.6**	**26.6**	**29.4**	**31.8**	**0.9**
North Africa	3.8	3.7	2.4	3.2	3.3	3.4	3.6	3.7	0.0
West Africa	4.1	4.5	4.4	4.4	4.7	4.8	5.0	5.0	0.4
South America	3.0	2.9	3.0	2.9	2.9	2.9	3.0	3.2	0.3
Non-OPEC	**50.4**	**51.8**	**52.0**	**61.7**	**65.3**	**68.6**	**72.7**	**78.4**	**1.4**
OECD	**21.0**	**21.4**	**21.5**	**27.9**	**29.1**	**29.4**	**29.9**	**30.4**	**1.2**
OECD North America	15.3	16.0	16.4	23.8	25.0	25.1	25.5	25.7	1.6
United States	8.9	9.3	9.7	15.3	15.2	14.2	13.8	13.5	1.2
Canada	3.4	3.6	3.7	5.4	6.4	7.3	7.8	8.0	2.7
Mexico and Chile	3.0	3.0	3.0	3.1	3.4	3.7	3.9	4.2	1.1
OECD Europe	4.9	4.6	4.3	3.3	3.2	3.2	3.2	3.4	-1.0
North Sea	3.9	3.6	3.3	2.2	2.1	2.0	2.0	2.1	-1.7
Other	1.0	1.0	1.0	1.0	1.1	1.2	1.2	1.3	1.0
OECD Asia	**0.8**	**0.8**	**0.8**	**0.8**	**1.0**	**1.1**	**1.2**	**1.2**	**1.2**
Australia and New Zealand	0.7	0.7	0.6	0.6	0.8	0.9	1.0	1.0	1.4
Other	0.2	0.2	0.2	0.2	0.2	0.2	0.2	0.2	0.2
Non-OECD	**29.4**	**30.5**	**30.5**	**33.8**	**36.2**	**39.2**	**42.9**	**48.0**	**1.5**
Non-OECD Europe and Eurasia	**13.1**	**13.4**	**13.5**	**14.8**	**15.4**	**16.6**	**17.8**	**19.1**	**1.2**
Russia	**9.9**	**10.1**	**10.2**	**11.0**	**11.1**	**11.6**	**12.2**	**12.7**	**0.7**
Caspian Area	**2.8**	**3.0**	**3.0**	**3.6**	**4.1**	**4.7**	**5.3**	**6.2**	**2.5**
Kazakhstan	1.5	1.6	1.6	2.3	2.9	3.5	4.0	4.6	3.5
Other	1.3	1.3	1.3	1.3	1.2	1.3	1.3	1.5	0.5
Other	**0.3**	**0.3**	**0.3**	**0.2**	**0.2**	**0.2**	**0.2**	**0.3**	**-0.6**
Non-OECD Asia	**7.8**	**8.2**	**8.2**	**9.0**	**9.7**	**9.9**	**10.5**	**11.6**	**1.2**
China	4.1	4.4	4.3	5.2	6.0	6.1	6.7	7.5	1.8
India	0.9	1.0	1.0	1.0	1.1	1.1	1.2	1.4	1.2
Other	2.9	2.9	2.9	2.8	2.7	2.6	2.6	2.7	-0.2
Middle East (Non-OPEC)	**1.6**	**1.6**	**1.5**	**1.1**	**1.1**	**1.0**	**1.0**	**1.0**	**-1.6**
Oman	0.8	0.9	0.9	0.7	0.6	0.6	0.5	0.5	-1.7
Other	0.7	0.7	0.6	0.4	0.5	0.5	0.5	0.5	-1.6
Africa	**2.6**	**2.6**	**2.6**	**2.8**	**2.8**	**2.9**	**3.0**	**3.4**	**0.9**
Ghana	0.0	0.0	0.1	0.3	0.3	0.3	0.3	0.3	12.8
Other	2.6	2.6	2.6	2.5	2.5	2.6	2.7	3.1	0.6
Central and South America	**4.3**	**4.6**	**4.7**	**6.1**	**7.1**	**8.8**	**10.6**	**12.9**	**3.5**
Brazil	2.4	2.5	2.5	3.5	4.5	5.6	6.8	8.2	4.0
Other	1.9	2.1	2.1	2.6	2.6	3.2	3.8	4.7	2.8
Total World	84.5	87.2	87.7	94.8	99.8	106.4	113.8	122.1	1.1
OPEC Share of World Production	40%	41%	41%	35%	35%	36%	36%	36%	
Persian Gulf Share of World Production	27%	28%	30%	24%	24%	25%	26%	26%	

[a]OPEC = Organization of the Petroleum Exporting Countries (OPEC-13).
Note: Totals may not equal sum of components due to independent rounding.
Sources: **History:** U.S. Energy Information Administration (EIA), Office of Energy Analysis and Office of Petroleum, Natural Gas & Biofuels Analysis. **Projections:** EIA, Generate World Oil Balance application (2014), run IEO2014_GWOB_HighPriceCase.xlsx.

Table B5. World crude and lease condensate[a] production by region and country, High Oil Price case, 2009-40
(million barrels per day)

Region	History			Projections					Average annual percent change, 2010-40
	2009	2010	2011	2020	2025	2030	2035	2040	
OPEC[b]	**31.0**	**32.0**	**32.2**	**28.5**	**29.6**	**32.5**	**35.2**	**37.9**	**0.6**
Middle East	20.8	21.7	23.0	19.1	19.9	22.6	25.2	27.5	0.8
North Africa	3.3	3.2	2.0	2.6	2.6	2.7	2.7	2.7	-0.6
West Africa	4.1	4.4	4.3	4.3	4.5	4.6	4.8	4.8	0.3
South America	2.8	2.7	2.8	2.5	2.5	2.5	2.6	2.8	0.2
Non-OPEC	**41.9**	**42.9**	**42.8**	**50.7**	**52.6**	**55.0**	**57.6**	**61.3**	**1.2**
OECD	**15.3**	**15.4**	**15.2**	**21.0**	**21.9**	**22.0**	**22.1**	**22.3**	**1.3**
OECD North America	**10.8**	**11.2**	**11.5**	**18.5**	**19.3**	**19.4**	**19.5**	**19.6**	**1.9**
United States	5.5	5.6	5.8	11.0	10.6	9.6	8.9	8.5	1.4
Canada	2.6	2.9	3.0	4.8	5.7	6.5	7.0	7.2	3.1
Mexico and Chile	2.7	2.6	2.6	2.7	3.0	3.3	3.6	3.8	1.2
OECD Europe	**3.9**	**3.6**	**3.3**	**2.1**	**1.9**	**1.9**	**1.8**	**1.9**	**-2.1**
North Sea	3.4	3.1	2.8	1.7	1.6	1.5	1.5	1.6	-2.1
Other	0.5	0.5	0.5	0.4	0.3	0.3	0.3	0.3	-2.0
OECD Asia	**0.5**	**0.5**	**0.5**	**0.5**	**0.7**	**0.7**	**0.8**	**0.8**	**1.5**
Australia and New Zealand	0.5	0.5	0.5	0.5	0.7	0.7	0.8	0.8	1.5
Other	0.0	0.0	0.0	0.0	0.0	0.0	0.0	0.0	1.2
Non-OECD	**26.6**	**27.5**	**27.5**	**29.7**	**30.7**	**33.1**	**35.5**	**39.0**	**1.2**
Non-OECD Europe and Eurasia	**12.4**	**12.7**	**12.8**	**13.9**	**14.3**	**15.2**	**16.2**	**17.1**	**1.0**
Russia	9.5	9.7	9.8	10.4	10.4	10.7	11.2	11.4	0.5
Caspian Area	**2.7**	**2.8**	**2.8**	**3.4**	**3.8**	**4.4**	**4.8**	**5.5**	**2.3**
Kazakhstan	1.5	1.6	1.6	2.2	2.7	3.2	3.7	4.1	3.3
Other	1.2	1.3	1.2	1.2	1.1	1.2	1.2	1.4	0.4
Other	**0.3**	**0.2**	**0.2**	**0.2**	**0.2**	**0.2**	**0.2**	**0.2**	**-1.0**
Non-OECD Asia	**6.9**	**7.3**	**7.2**	**7.5**	**7.4**	**7.3**	**7.2**	**7.2**	**0.0**
China	3.8	4.1	4.1	4.6	4.7	4.7	4.6	4.5	0.3
India	0.7	0.7	0.8	0.6	0.6	0.6	0.7	0.8	0.1
Other	2.4	2.4	2.3	2.2	2.1	2.0	1.9	1.9	-0.8
Middle East (Non-OPEC)	**1.5**	**1.5**	**1.5**	**1.1**	**1.1**	**1.0**	**0.9**	**0.9**	**-1.7**
Africa	**2.2**	**2.2**	**2.2**	**2.2**	**2.2**	**2.3**	**2.3**	**2.7**	**0.7**
Central and South America	**3.6**	**3.8**	**3.9**	**5.0**	**5.7**	**7.2**	**8.9**	**11.0**	**3.7**
Brazil	2.0	2.1	2.1	2.8	3.6	4.5	5.5	6.8	4.1
Other	1.6	1.7	1.8	2.1	2.2	2.8	3.3	4.2	3.0
Total World	**72.9**	**74.9**	**75.0**	**79.3**	**82.2**	**87.5**	**92.8**	**99.1**	**0.9**
OPEC Share of World Production	42%	43%	43%	36%	36%	37%	38%	38%	
Persian Gulf Share of World Production	29%	29%	31%	24%	24%	26%	27%	28%	

[a]Crude and lease condensate includes tight oil, shale oil, extra-heavy oil, field condensate, and bitumen.
[b]OPEC = Organization of the Petroleum Exporting Countries (OPEC-13).
Note: Totals may not equal sum of components due to independent rounding.
Sources: **History:** U.S. Energy Information Administration (EIA), Office of Energy Analysis and Office of Petroleum, Natural Gas & Biofuels Analysis. **Projections:** EIA, Generate World Oil Balance application (2014), run IEO2014_GWOB_HighPriceCase.xlsx.

Table B6. World other liquid fuels[a] production by region and country, High Oil Price case, 2009-40
(million barrels per day)

Region	History			Projections					Average annual percent change, 2010-40
	2009	2010	2011	2020	2025	2030	2035	2040	
OPEC[b]	**3.1**	**3.3**	**3.5**	**4.6**	**4.9**	**5.3**	**5.8**	**5.9**	**1.9**
Natural gas plant liquids	3.1	3.3	3.4	4.3	4.6	4.9	5.3	5.3	1.6
Biofuels[c]	0.0	0.0	0.0	0.0	0.0	0.0	0.0	0.0	—
Coal-to-liquids	0.0	0.0	0.0	0.0	0.0	0.0	0.0	0.0	—
Gas-to-liquids (primarily Qatar)	0.0	0.0	0.1	0.3	0.3	0.4	0.4	0.5	14.6
Refinery gain	0.0	0.0	0.0	0.0	0.0	0.0	0.0	0.1	1.0
Non-OPEC	**8.5**	**9.0**	**9.3**	**11.0**	**12.6**	**13.5**	**15.1**	**17.1**	**2.3**
OECD	**5.6**	**6.0**	**6.3**	**6.8**	**7.2**	**7.4**	**7.8**	**8.1**	**1.1**
Natural gas plant liquids	3.4	3.5	3.6	4.1	4.4	4.5	4.6	4.5	0.8
Biofuels[c]	0.7	0.8	0.9	1.1	1.2	1.3	1.4	1.5	2.0
Coal-to-liquids	0.0	0.0	0.0	0.0	0.0	0.0	0.0	0.2	19.4
Gas-to-liquids	0.0	0.0	0.0	0.0	0.1	0.1	0.2	0.3	—
Kerogen	0.0	0.0	0.0	0.0	0.0	0.0	0.0	0.0	0.8
Refinery gain	1.6	1.7	1.7	1.6	1.5	1.5	1.6	1.6	0.1
Non-OECD	**2.7**	**2.9**	**3.0**	**4.1**	**5.4**	**6.1**	**7.4**	**9.0**	**4.1**
Natural gas plant liquids	1.6	1.6	1.7	2.1	2.4	2.6	2.9	3.3	2.4
Biofuels[c]	0.4	0.5	0.5	0.7	1.0	1.3	1.6	2.0	4.7
Coal-to-liquids	0.2	0.2	0.2	0.5	1.1	1.1	1.7	2.5	9.3
Gas-to-liquids	0.1	0.1	0.1	0.1	0.1	0.1	0.1	0.1	2.5
Refinery gain	0.5	0.6	0.6	0.8	0.9	1.0	1.1	1.2	3.8
Total World	**11.6**	**12.3**	**12.8**	**15.5**	**17.6**	**18.8**	**20.9**	**22.9**	**2.1**
Natural Gas Plant Liquids	**8.1**	**8.4**	**8.7**	**10.5**	**11.4**	**12.0**	**12.8**	**13.1**	**1.5**
United States	1.9	2.1	2.2	2.7	3.0	3.1	3.3	3.1	1.4
Russia	0.4	0.4	0.4	0.6	0.7	0.8	1.0	1.3	3.6
Biofuels[c]	**1.2**	**1.3**	**1.4**	**1.8**	**2.2**	**2.6**	**3.0**	**3.4**	**3.3**
Brazil	0.3	0.3	0.3	0.4	0.6	0.7	0.9	1.0	3.6
China	0.0	0.0	0.0	0.1	0.1	0.2	0.3	0.4	9.9
India	0.0	0.0	0.0	0.0	0.0	0.0	0.0	0.0	8.5
United States	0.5	0.6	0.6	0.7	0.7	0.7	0.7	0.7	0.6
Coal-to-liquids	**0.2**	**0.2**	**0.2**	**0.5**	**1.1**	**1.2**	**1.7**	**2.7**	**9.6**
Australia/New Zealand	0.0	0.0	0.0	0.0	0.0	0.0	0.0	0.0	—
China	0.0	0.0	0.0	0.2	0.7	0.7	1.3	2.0	19.4
Germany	0.0	0.0	0.0	0.0	0.0	0.0	0.0	0.0	0.5
India	0.0	0.0	0.0	0.0	0.1	0.1	0.1	0.1	—
South Africa	0.2	0.2	0.2	0.3	0.3	0.3	0.3	0.3	2.0
United States	0.0	0.0	0.0	0.0	0.0	0.0	0.0	0.2	—
Gas-to-liquids	**0.1**	**0.1**	**0.1**	**0.3**	**0.5**	**0.6**	**0.8**	**0.9**	**9.0**
Qatar	0.0	0.0	0.1	0.2	0.3	0.4	0.4	0.4	—
South Africa	0.0	0.0	0.0	0.0	0.0	0.1	0.1	0.1	—
Refinery Gain	**2.2**	**2.3**	**2.4**	**2.4**	**2.4**	**2.6**	**2.7**	**2.8**	**0.7**
United States	1.0	1.1	1.1	1.0	0.9	0.8	0.8	0.8	-0.8
China	0.2	0.2	0.2	0.3	0.4	0.4	0.5	0.5	2.7

[a]Other liquid fuels include natural gas plant liquids, biofuels, gas-to-liquids, coal-to-liquids, kerogen, and refinery gain.
[b]OPEC = Organization of the Petroleum Exporting Countries (OPEC-13).
[c]Ethanol values are reported on a gasoline-equivalent basis.
Note: Totals may not equal sum of components due to independent rounding.
Sources: **History:** U.S. Energy Information Administration (EIA), Office of Energy Analysis and Office of Petroleum, Natural Gas & Biofuels Analysis. **Projections:** EIA, Generate World Oil Balance application (2014), run IEO2014_GWOB_HighPriceCase.xlsx.

THIS PAGE INTENTIONALLY LEFT BLANK

Appendix C
Low Oil Price case projections

THIS PAGE INTENTIONALLY LEFT BLANK

Table C1. World gross domestic product (GDP) by region expressed in purchasing power parity, Low Oil Price case, 2009-40
(billion 2005 dollars)

Region	History		Projections					Average annual percent change, 2010-40
	2009	2010	2020	2025	2030	2035	2040	
OECD								
OECD Americas	**15,498**	**15,929**	**20,692**	**23,228**	**26,077**	**29,200**	**32,663**	**2.4**
United States[a]	12,758	13,063	16,739	18,750	21,150	23,794	26,725	2.4
Canada	1,165	1,202	1,534	1,714	1,894	2,085	2,284	2.2
Mexico/Chile	1,575	1,664	2,419	2,764	3,033	3,321	3,654	2.7
OECD Europe	**14,262**	**14,618**	**17,778**	**19,836**	**21,922**	**24,137**	**26,656**	**2.0**
OECD Asia	**5,791**	**6,062**	**7,361**	**7,965**	**8,462**	**8,867**	**9,305**	**1.4**
Japan	3,776	3,948	4,477	4,685	4,812	4,874	4,861	0.7
South Korea	1,244	1,323	1,849	2,117	2,354	2,550	2,836	2.6
Australia/NewZealand	771	790	1,035	1,163	1,296	1,443	1,607	2.4
Total OECD	**35,551**	**36,609**	**45,831**	**51,029**	**56,461**	**62,204**	**68,623**	**2.1**
Non-OECD								
Non-OECD Europe and Eurasia	**4,346**	**4,502**	**6,070**	**7,088**	**8,092**	**9,194**	**10,190**	**2.8**
Russia	1,938	2,022	2,625	2,996	3,324	3,669	3,890	2.2
Other	2,408	2,480	3,445	4,092	4,768	5,525	6,300	3.2
Non-OECD Asia	**16,628**	**18,206**	**32,663**	**42,000**	**52,024**	**61,944**	**71,299**	**4.7**
China	8,299	9,167	17,451	22,089	27,292	31,822	34,868	4.6
India	3,364	3,661	6,310	8,428	10,477	12,706	15,244	4.9
Other	4,965	5,379	8,902	11,483	14,256	17,416	21,187	4.7
Middle East	**2,263**	**2,292**	**3,463**	**4,228**	**5,045**	**5,959**	**6,967**	**3.8**
Africa	**3,780**	**3,963**	**6,193**	**7,787**	**9,828**	**12,391**	**15,665**	**4.7**
Central and South America	**4,623**	**4,927**	**6,841**	**8,085**	**9,452**	**10,993**	**12,890**	**3.3**
Brazil	1,833	1,971	2,624	3,102	3,645	4,281	5,101	3.2
Other	2,790	2,955	4,217	4,983	5,807	6,713	7,789	3.3
Total Non-OECD	**31,640**	**33,889**	**55,231**	**69,188**	**84,442**	**100,481**	**117,012**	**4.2**
Total World	**67,192**	**70,498**	**101,062**	**120,217**	**140,903**	**162,685**	**185,635**	**3.3**

[a] Includes the 50 States and the District of Columbia.
Note: Totals may not equal sum of components due to independent rounding.
Sources: Derived from Oxford Economic Model (February 2014), www.oxfordeconomics.com (subscription site); EIA, *Annual Energy Outlook 2014*, DOE/EIA-0383(2014) (Washington, DC: April 2014); and AEO2014 National Energy Modeling System, run LOWPRICE.D120613A (Low Oil Price case), www.eia.gov/aeo.

Table C2. World liquids consumption by region, Low Oil Price case, 2009-40
(million barrels per day)

Region	History		Projections					Average annual percent change, 2010-40
	2009	2010	2020	2025	2030	2035	2040	
OECD								
OECD Americas	**23.1**	**23.5**	**24.8**	**24.8**	**24.7**	**24.9**	**25.5**	**0.3**
United States[a]	18.6	18.9	19.6	19.5	19.4	19.4	19.8	0.1
Canada	2.2	2.2	2.4	2.4	2.4	2.5	2.6	0.5
Mexico/Chile	2.4	2.4	2.8	2.9	3.0	3.0	3.2	1.0
OECD Europe	**15.0**	**14.8**	**14.7**	**15.0**	**15.2**	**15.4**	**15.9**	**0.2**
OECD Asia	**7.7**	**7.7**	**8.4**	**8.4**	**8.3**	**8.2**	**8.1**	**0.2**
Japan	4.4	4.4	4.5	4.5	4.4	4.2	4.0	-0.3
South Korea	2.2	2.3	2.7	2.8	2.8	2.8	2.8	0.7
Australia/NewZealand	1.1	1.1	1.2	1.2	1.2	1.2	1.3	0.4
Total OECD	**45.8**	**46.0**	**47.8**	**48.3**	**48.3**	**48.5**	**49.6**	**0.2**
Non-OECD								
Non-OECD Europe and Eurasia	**4.8**	**4.8**	**5.5**	**5.6**	**5.8**	**5.9**	**6.0**	**0.7**
Russia	3.0	3.0	3.3	3.3	3.3	3.3	3.2	0.2
Other	1.8	1.8	2.2	2.3	2.5	2.6	2.8	1.4
Non-OECD Asia	**18.4**	**19.8**	**26.5**	**29.9**	**33.6**	**36.7**	**39.5**	**2.3**
China	8.5	9.3	13.0	14.3	15.9	16.9	17.2	2.1
India	3.1	3.3	4.3	4.9	5.4	5.8	6.1	2.1
Other	6.7	7.2	9.1	10.8	12.4	14.1	16.2	2.7
Middle East	**6.5**	**6.7**	**8.5**	**9.1**	**10.2**	**11.3**	**12.5**	**2.1**
Africa	**3.3**	**3.4**	**4.0**	**4.4**	**4.9**	**5.6**	**6.4**	**2.2**
Central and South America	**5.7**	**6.0**	**7.1**	**7.3**	**7.8**	**8.4**	**9.2**	**1.4**
Brazil	2.5	2.6	3.3	3.4	3.7	4.0	4.5	1.8
Other	3.3	3.4	3.8	3.9	4.2	4.4	4.7	1.1
Total Non-OECD	**38.7**	**40.7**	**51.6**	**56.4**	**62.3**	**67.8**	**73.7**	**2.0**
Total World	**84.5**	**86.8**	**99.4**	**104.6**	**110.6**	**116.4**	**123.3**	**1.2**

[a]Includes the 50 States and the District of Columbia.
Note: Totals may not equal sum of components due to independent rounding.
Sources: **History:** U.S. Energy Information Administration (EIA), International Energy Statistics database (as of November 2013), www.eia.gov/ies.
Projections: EIA, *Annual Energy Outlook 2014*, DOE/EIA-0383(2014) (Washington, DC: April 2014), AEO2014 National Energy Modeling System, run LOWPRICE.D120613A, www.eia.gov/aeo; and World Energy Projection System Plus (2014), run 2014.03.24_145137 (Low Oil Price case).

Table C3. World petroleum and other liquids consumption by region and end-use sector, Low Oil Price case, 2010-40 (quadrillion Btu)

Region	History 2010	Projections 2020	Projections 2025	Projections 2030	Projections 2035	Projections 2040	Average annual percent change, 2010-40
OECD							
United States							
Residential	1.1	0.9	0.8	0.8	0.7	0.7	-1.6
Commercial	0.7	0.7	0.7	0.8	0.8	0.8	0.7
Industrial	8.1	9.7	10.1	10.2	10.2	10.3	0.8
Transportation	26.9	26.2	25.6	25.3	25.5	26.1	-0.1
Electricity	0.4	0.2	0.2	0.2	0.2	0.2	-2.3
Total	**37.2**	**37.6**	**37.4**	**37.2**	**37.4**	**38.2**	**0.1**
Canada							
Residential	0.1	0.1	0.1	0.1	0.1	0.1	0.2
Commercial	0.1	0.1	0.1	0.1	0.1	0.1	0.5
Industrial	1.7	1.9	2.0	2.2	2.3	2.4	1.2
Transportation	2.4	2.6	2.5	2.4	2.4	2.5	0.1
Electricity	0.0	0.0	0.0	0.0	0.0	0.0	-1.0
Total	**4.3**	**4.8**	**4.8**	**4.8**	**5.0**	**5.2**	**0.6**
Mexico/Chile							
Residential	0.3	0.3	0.4	0.4	0.4	0.4	0.7
Commercial	0.1	0.1	0.1	0.1	0.1	0.1	0.9
Industrial	1.1	1.4	1.5	1.7	1.8	1.9	1.7
Transportation	2.7	3.2	3.3	3.3	3.3	3.5	0.9
Electricity	0.4	0.4	0.4	0.4	0.4	0.3	-0.7
Total	**4.7**	**5.4**	**5.7**	**5.8**	**5.9**	**6.2**	**1.0**
OECD Europe							
Residential	2.1	2.0	2.0	2.0	1.9	1.9	-0.2
Commercial	0.9	0.9	0.9	0.9	0.8	0.8	-0.3
Industrial	9.6	10.2	10.9	11.5	12.1	12.7	1.0
Transportation	17.7	17.0	17.0	16.9	16.8	17.3	-0.1
Electricity	0.4	0.4	0.4	0.4	0.3	0.3	-0.7
Total	**30.6**	**30.4**	**31.1**	**31.6**	**32.0**	**33.1**	**0.3**
Japan							
Residential	0.6	0.5	0.5	0.5	0.5	0.5	-0.6
Commercial	0.7	0.6	0.7	0.7	0.6	0.6	-0.2
Industrial	3.6	4.2	4.3	4.3	4.2	4.0	0.4
Transportation	3.7	3.3	3.1	3.0	2.9	2.8	-0.9
Electricity	0.5	0.5	0.4	0.4	0.4	0.4	-1.0
Total	**9.0**	**9.1**	**9.1**	**8.9**	**8.6**	**8.3**	**-0.3**

See notes at end of table.

Table C3. World petroleum and other liquids consumption by region and end-use sector, Low Oil Price case, 2010-40 (continued) (quadrillion Btu)

Region	History 2010	Projections 2020	2025	2030	2035	2040	Average annual percent change, 2010-40
OECD (continued)							
South Korea							
Residential	0.1	0.1	0.1	0.1	0.1	0.2	0.5
Commercial	0.1	0.1	0.1	0.1	0.1	0.1	0.0
Industrial	2.5	3.0	3.1	3.1	3.0	2.9	0.6
Transportation	1.8	2.2	2.3	2.3	2.4	2.5	1.2
Electricity	0.1	0.1	0.1	0.1	0.1	0.1	-1.0
Total	**4.6**	**5.6**	**5.7**	**5.8**	**5.7**	**5.8**	**0.8**
Australia/New Zealand							
Residential	0.0	0.0	0.0	0.0	0.0	0.0	0.3
Commercial	0.0	0.0	0.0	0.0	0.0	0.0	0.3
Industrial	0.6	0.6	0.6	0.6	0.7	0.7	0.5
Transportation	1.6	1.7	1.7	1.7	1.7	1.8	0.5
Electricity	0.0	0.0	0.0	0.0	0.0	0.0	-0.9
Total	**2.2**	**2.4**	**2.4**	**2.4**	**2.4**	**2.6**	**0.5**
Total OECD							
Residential	4.3	4.0	4.0	3.9	3.8	3.8	-0.5
Commercial	2.6	2.6	2.7	2.7	2.7	2.7	0.1
Industrial	27.2	31.0	32.6	33.5	34.1	35.0	0.8
Transportation	56.6	56.0	55.4	54.9	55.0	56.5	0.0
Electricity	1.9	1.6	1.5	1.5	1.4	1.4	-1.1
Total	**92.6**	**95.3**	**96.2**	**96.4**	**97.0**	**99.3**	**0.2**
Non-OECD							
Russia							
Residential	0.3	0.3	0.3	0.3	0.3	0.3	0.0
Commercial	0.1	0.1	0.1	0.1	0.1	0.1	-0.9
Industrial	2.7	2.8	2.8	2.8	2.9	2.7	0.1
Transportation	2.9	3.4	3.4	3.4	3.3	3.3	0.5
Electricity	0.1	0.1	0.1	0.1	0.1	0.1	-0.6
Total	**6.0**	**6.7**	**6.6**	**6.7**	**6.7**	**6.5**	**0.2**
Other Non-OECD Europe and Eurasia							
Residential	0.1	0.1	0.1	0.1	0.1	0.1	0.3
Commercial	0.1	0.1	0.1	0.1	0.1	0.1	0.0
Industrial	1.4	1.3	1.3	1.3	1.4	1.4	0.0
Transportation	2.0	2.9	3.1	3.4	3.7	4.0	2.4
Electricity	0.1	0.1	0.1	0.1	0.1	0.1	-0.9
Total	**3.7**	**4.6**	**4.8**	**5.1**	**5.4**	**5.8**	**1.5**

See notes at end of table.

Table C3. World petroleum and other liquids consumption by region and end-use sector, Low Oil Price case, 2010-40 (continued) (quadrillion Btu)

Region	History 2010	Projections 2020	2025	2030	2035	2040	Average annual percent change, 2010-40
Non-OECD (continued)							
China							
Residential	1.2	1.2	1.2	1.1	1.1	1.0	-0.6
Commercial	1.1	1.1	1.1	1.1	1.0	0.9	-0.4
Industrial	8.4	9.7	10.1	11.0	11.6	11.7	1.1
Transportation	8.4	14.5	16.8	19.2	20.9	21.8	3.2
Electricity	0.1	0.1	0.0	0.0	0.0	0.0	-4.8
Total	**19.1**	**26.6**	**29.2**	**32.5**	**34.6**	**35.5**	**2.1**
India							
Residential	0.9	1.2	1.1	1.1	1.1	1.0	0.3
Commercial	0.0	0.0	0.0	0.0	0.0	0.0	0.0
Industrial	3.2	3.6	4.0	4.5	4.9	4.9	1.5
Transportation	2.3	3.7	4.5	5.1	5.7	6.4	3.5
Electricity	0.2	0.2	0.2	0.2	0.2	0.2	-0.8
Total	**6.6**	**8.7**	**9.9**	**11.0**	**11.8**	**12.5**	**2.1**
Other Non-OECD Asia							
Residential	0.5	0.6	0.6	0.7	0.7	0.8	1.2
Commercial	0.3	0.3	0.4	0.4	0.5	0.5	1.7
Industrial	4.8	5.9	7.2	8.5	9.9	11.5	3.0
Transportation	8.2	11.1	13.1	15.1	17.2	20.0	3.0
Electricity	1.1	1.1	1.0	1.0	0.9	0.9	-0.6
Total	**14.9**	**19.0**	**22.4**	**25.7**	**29.2**	**33.7**	**2.8**
Middle East							
Residential	0.7	0.8	0.7	0.7	0.7	0.7	0.1
Commercial	0.1	0.1	0.1	0.2	0.2	0.2	1.1
Industrial	3.8	5.4	6.4	8.2	10.2	12.3	4.0
Transportation	5.8	7.6	8.0	8.4	8.6	9.0	1.5
Electricity	3.4	3.4	3.3	3.1	3.0	2.8	-0.6
Total	**13.8**	**17.3**	**18.5**	**20.6**	**22.6**	**25.0**	**2.0**
Africa							
Residential	0.7	0.8	0.8	0.9	0.9	1.0	1.2
Commercial	0.1	0.1	0.1	0.1	0.1	0.2	2.0
Industrial	1.5	1.6	1.8	2.0	2.2	2.5	1.6
Transportation	3.7	4.8	5.5	6.3	7.3	8.7	2.9
Electricity	0.8	0.8	0.7	0.7	0.7	0.6	-0.8
Total	**6.9**	**8.1**	**8.9**	**10.0**	**11.3**	**13.0**	**2.2**

See notes at end of table.

Table C3. **World petroleum and other liquids consumption by region and end-use sector, Low Oil Price case, 2010-40 (continued)** (quadrillion Btu)

	History	Projections					Average annual percent change, 2010-40
Region	2010	2020	2025	2030	2035	2040	
Non-OECD (continued)							
Brazil							
Residential	0.3	0.3	0.3	0.4	0.4	0.4	0.8
Commercial	0.0	0.0	0.0	0.0	0.0	0.0	1.1
Industrial	2.0	2.4	2.4	2.6	3.0	3.4	1.7
Transportation	2.9	3.8	4.1	4.4	4.7	5.3	2.1
Electricity	0.1	0.1	0.1	0.1	0.1	0.1	-0.9
Total	**5.4**	**6.7**	**6.9**	**7.5**	**8.2**	**9.2**	**1.8**
Other Central and South America							
Residential	0.3	0.4	0.4	0.4	0.4	0.4	0.7
Commercial	0.1	0.1	0.1	0.1	0.1	0.1	1.4
Industrial	2.1	2.2	2.2	2.4	2.6	2.9	1.1
Transportation	3.4	4.3	4.6	4.8	5.1	5.5	1.6
Electricity	1.0	1.0	0.9	0.9	0.8	0.8	-0.9
Total	**6.9**	**7.8**	**8.1**	**8.5**	**9.0**	**9.7**	**1.1**
Total Non-OECD							
Residential	5.1	5.7	5.7	5.7	5.7	5.7	0.4
Commercial	1.9	2.0	2.0	2.1	2.1	2.1	0.4
Industrial	29.8	34.9	37.9	43.4	48.7	53.4	2.0
Transportation	39.6	56.1	63.2	70.1	76.4	83.9	2.5
Electricity	6.9	6.8	6.5	6.2	5.9	5.6	-0.7
Total	**83.3**	**105.4**	**115.3**	**127.5**	**138.8**	**150.8**	**2.0**
Total World							
Residential	9.5	9.7	9.7	9.6	9.6	9.5	0.0
Commercial	4.5	4.6	4.7	4.8	4.8	4.8	0.2
Industrial	57.0	65.9	70.5	76.9	82.8	88.4	1.5
Transportation	96.2	112.1	118.6	125.0	131.4	140.4	1.3
Electricity	8.8	8.3	8.0	7.6	7.3	7.0	-0.8
Total	**176.0**	**200.7**	**211.5**	**223.9**	**235.8**	**250.1**	**1.2**

Note: Totals may not equal sum of components due to independent rounding.
Sources: **2010:** Derived from U.S. Energy Information Administration (EIA), International Energy Statistics database (as of November 2013), www.eia.gov/ies; and International Energy Agency, "Balances of OECD and Non-OECD Statistics" (2013), www.iea.org (subscription site).
Projections: EIA, *Annual Energy Outlook 2014*, DOE/EIA-0383(2014) (Washington, DC: April 2014), AEO2014 National Energy Modeling System, run LOWPRICE.D120613A, www.eia.gov/aeo; and World Energy Projection System Plus (2014), run 2014.03.24_145137 (Low Oil Price case).

Table C4. World petroleum and other liquids production by region and country, Low Oil Price case, 2009-40
(million barrels per day)

Region	History			Projections					Average annual percent change, 2010-40
	2009	2010	2011	2020	2025	2030	2035	2040	
OPEC[a]	34.1	35.4	35.7	43.3	48.7	54.6	59.9	65.3	2.1
Middle East	23.2	24.3	25.9	30.4	34.5	38.9	43.0	47.3	2.2
North Africa	3.8	3.7	2.4	3.7	4.0	4.3	4.7	4.9	0.9
West Africa	4.1	4.5	4.4	5.5	6.2	6.8	7.2	7.5	1.7
South America	3.0	2.9	3.0	3.6	4.0	4.6	5.0	5.6	2.2
Non-OPEC	50.4	51.8	52.0	56.1	55.9	56.0	56.5	58.0	0.4
OECD	21.0	21.4	21.5	24.5	23.6	22.5	22.2	21.8	0.1
OECD North America	15.3	16.0	16.4	20.7	20.0	19.3	19.1	18.7	0.5
United States	8.9	9.3	9.7	13.6	12.9	11.7	11.2	10.6	0.4
Canada	3.4	3.6	3.7	4.7	5.1	5.5	5.7	5.8	1.6
Mexico and Chile	3.0	3.0	3.0	2.4	2.0	2.0	2.1	2.2	-1.0
OECD Europe	4.9	4.6	4.3	3.1	2.9	2.5	2.4	2.5	-2.0
North Sea	3.9	3.6	3.3	2.2	2.0	1.6	1.5	1.5	-2.9
Other	1.0	1.0	1.0	0.9	0.9	0.9	0.9	1.0	0.0
OECD Asia	0.8	0.8	0.8	0.7	0.7	0.7	0.7	0.7	-0.6
Australia and New Zealand	0.7	0.7	0.6	0.5	0.5	0.5	0.5	0.5	-0.9
Other	0.2	0.2	0.2	0.2	0.2	0.2	0.2	0.2	0.4
Non-OECD	29.4	30.5	30.5	31.6	32.3	33.5	34.3	36.2	0.6
Non-OECD Europe and Eurasia	13.1	13.4	13.5	14.1	14.6	15.0	15.5	16.2	0.6
Russia	9.9	10.1	10.2	10.6	10.6	10.8	11.1	11.3	0.4
Caspian Area	2.8	3.0	3.0	3.1	3.7	3.9	4.2	4.5	1.4
Kazakhstan	1.5	1.6	1.6	1.9	2.4	2.6	2.8	3.1	2.2
Other	1.3	1.3	1.3	1.2	1.3	1.4	1.4	1.4	0.2
Other	0.3	0.3	0.3	0.3	0.3	0.3	0.3	0.3	-0.2
Non-OECD Asia	7.8	8.2	8.2	8.4	8.4	8.6	8.5	8.6	0.2
China	4.1	4.4	4.3	4.8	4.9	5.1	5.0	5.0	0.5
India	0.9	1.0	1.0	1.0	1.0	1.1	1.1	1.2	0.7
Other	2.9	2.9	2.9	2.6	2.4	2.4	2.3	2.4	-0.6
Middle East (Non-OPEC)	1.6	1.6	1.5	1.0	0.9	0.7	0.6	0.6	-3.5
Oman	0.8	0.9	0.9	0.7	0.6	0.5	0.4	0.3	-3.7
Other	0.7	0.7	0.6	0.3	0.3	0.3	0.3	0.3	-3.2
Africa	2.6	2.6	2.6	2.7	2.8	2.9	2.9	3.2	0.7
Ghana	0.0	0.0	0.1	0.3	0.3	0.3	0.3	0.3	13.0
Other	2.6	2.6	2.6	2.4	2.5	2.6	2.6	2.9	0.4
Central and South America	4.3	4.6	4.7	5.3	5.7	6.3	6.8	7.6	1.7
Brazil	2.4	2.5	2.5	2.9	3.3	3.8	4.1	4.5	1.9
Other	1.9	2.1	2.1	2.5	2.4	2.5	2.7	3.1	1.4
Total World	84.5	87.2	87.7	99.4	104.6	110.6	116.4	123.3	1.2
OPEC Share of World Production	40%	41%	41%	44%	47%	49%	51%	53%	
Persian Gulf Share of World Production	27%	28%	30%	31%	33%	35%	37%	38%	

[a]OPEC = Organization of the Petroleum Exporting Countries (OPEC-13).
Note: Totals may not equal sum of components due to independent rounding.
Sources: **History:** U.S. Energy Information Administration (EIA), Office of Energy Analysis and Office of Petroleum, Natural Gas & Biofuels Analysis. **Projections:** EIA, Generate World Oil Balance application (2014), run IEO2014_GWOB_LowPriceCase.xlsx.

Table C5. World crude and lease condensate[a] production by region and country, Low Oil Price case, 2009-40
(million barrels per day)

Region	History			Projections					Average annual percent change, 2010-40
	2009	2010	2011	2020	2025	2030	2035	2040	
OPEC[b]	**31.0**	**32.0**	**32.2**	**39.0**	**44.2**	**49.9**	**54.8**	**60.2**	**2.1**
Middle East	20.8	21.7	23.0	27.1	31.0	35.3	39.3	43.6	2.3
North Africa	3.3	3.2	2.0	3.2	3.4	3.6	3.8	4.1	0.8
West Africa	4.1	4.4	4.3	5.4	6.0	6.7	7.0	7.3	1.7
South America	2.8	2.7	2.8	3.3	3.7	4.2	4.6	5.2	2.3
Non-OPEC	**41.9**	**42.9**	**42.8**	**45.5**	**44.4**	**43.8**	**43.7**	**44.7**	**0.1**
OECD	**15.3**	**15.4**	**15.2**	**17.8**	**16.6**	**15.5**	**15.0**	**14.7**	**-0.1**
OECD North America	10.8	11.2	11.5	15.3	14.4	13.6	13.4	13.1	0.5
United States	5.5	5.6	5.8	9.2	8.3	7.2	6.6	6.1	0.3
Canada	2.6	2.9	3.0	4.0	4.4	4.8	5.0	5.1	1.9
Mexico and Chile	2.7	2.6	2.6	2.1	1.7	1.7	1.8	1.9	-1.1
OECD Europe	**3.9**	**3.6**	**3.3**	**2.1**	**1.8**	**1.4**	**1.3**	**1.3**	**-3.4**
North Sea	3.4	3.1	2.8	1.7	1.5	1.2	1.0	1.0	-3.7
Other	0.5	0.5	0.5	0.4	0.3	0.3	0.3	0.3	-2.1
OECD Asia	**0.5**	**0.5**	**0.5**	**0.4**	**0.4**	**0.4**	**0.4**	**0.4**	**-1.3**
Australia and New Zealand	0.5	0.5	0.5	0.4	0.4	0.4	0.4	0.4	-1.3
Other	0.0	0.0	0.0	0.0	0.0	0.0	0.0	0.0	-1.6
Non-OECD	**26.6**	**27.5**	**27.5**	**27.6**	**27.8**	**28.4**	**28.7**	**29.9**	**0.3**
Non-OECD Europe and Eurasia	12.4	12.7	12.8	13.2	13.5	13.7	14.0	14.2	0.4
Russia	9.5	9.7	9.8	10.0	9.9	10.0	10.1	10.1	0.1
Caspian Area	**2.7**	**2.8**	**2.8**	**2.9**	**3.4**	**3.6**	**3.7**	**4.0**	**1.1**
Kazakhstan	1.5	1.6	1.6	1.8	2.2	2.3	2.5	2.6	1.8
Other	1.2	1.3	1.2	1.1	1.2	1.3	1.3	1.3	0.1
Other	**0.3**	**0.2**	**0.2**	**0.2**	**0.2**	**0.2**	**0.2**	**0.2**	**-0.5**
Non-OECD Asia	**6.9**	**7.3**	**7.2**	**7.0**	**6.8**	**6.7**	**6.5**	**6.4**	**-0.4**
China	3.8	4.1	4.1	4.2	4.3	4.3	4.2	4.0	0.0
India	0.7	0.7	0.8	0.7	0.6	0.6	0.6	0.7	-0.5
Other	2.4	2.4	2.3	2.1	1.9	1.8	1.7	1.8	-1.1
Middle East (Non-OPEC)	**1.5**	**1.5**	**1.5**	**1.0**	**0.8**	**0.7**	**0.6**	**0.5**	**-3.6**
Africa	**2.2**	**2.2**	**2.2**	**2.2**	**2.3**	**2.4**	**2.4**	**2.7**	**0.7**
Central and South America	**3.6**	**3.8**	**3.9**	**4.2**	**4.3**	**4.8**	**5.3**	**6.0**	**1.6**
Brazil	2.0	2.1	2.1	2.2	2.4	2.8	3.0	3.4	1.6
Other	1.6	1.7	1.8	2.1	1.9	2.1	2.2	2.6	1.5
Total World	**72.9**	**74.9**	**75.0**	**84.4**	**88.6**	**93.7**	**98.5**	**104.9**	**1.1**
OPEC Share of World Production	42%	43%	43%	46%	50%	53%	56%	57%	
Persian Gulf Share of World Production	29%	29%	31%	32%	35%	38%	40%	42%	

[a]Crude and lease condensate includes tight oil, shale oil, extra-heavy oil, field condensate, and bitumen.
[b]OPEC = Organization of the Petroleum Exporting Countries (OPEC-13).
Note: Totals may not equal sum of components due to independent rounding.
Sources: **History:** U.S. Energy Information Administration (EIA), Office of Energy Analysis and Office of Petroleum, Natural Gas & Biofuels Analysis. **Projections:** EIA, Generate World Oil Balance application (2014), run IEO2014_GWOB_LowPriceCase.xlsx.

Table C6. World other liquid fuels[a] production by region and country, Low Oil Price case, 2009-40
(million barrels per day)

Region	History			Projections					Average annual percent change, 2010-40
	2009	2010	2011	2020	2025	2030	2035	2040	
OPEC[b]	**3.1**	**3.3**	**3.5**	**4.3**	**4.5**	**4.8**	**5.1**	**5.0**	**1.4**
Natural gas plant liquids	3.1	3.3	3.4	4.0	4.2	4.4	4.8	4.7	1.2
Biofuels[c]	0.0	0.0	0.0	0.0	0.0	0.0	0.0	0.0	—
Coal-to-liquids	0.0	0.0	0.0	0.0	0.0	0.0	0.0	0.0	—
Gas-to-liquids (primarily Qatar)	0.0	0.0	0.1	0.3	0.3	0.3	0.3	0.3	12.3
Refinery gain	0.0	0.0	0.0	0.0	0.0	0.1	0.1	0.1	1.3
Non-OPEC	**8.5**	**9.0**	**9.3**	**10.7**	**11.5**	**12.1**	**12.8**	**13.4**	**1.3**
OECD	**5.8**	**6.0**	**6.3**	**6.7**	**7.0**	**7.0**	**7.2**	**7.1**	**0.5**
Natural gas plant liquids	3.4	3.5	3.6	4.0	4.1	4.1	4.2	4.1	0.5
Biofuels[c]	0.7	0.8	0.9	1.0	1.1	1.1	1.1	1.1	1.0
Coal-to-liquids	0.0	0.0	0.0	0.0	0.0	0.0	0.0	0.0	-0.1
Gas-to-liquids	0.0	0.0	0.0	0.0	0.0	0.0	0.0	0.0	—
Kerogen	0.0	0.0	0.0	0.0	0.0	0.0	0.0	0.0	-1.0
Refinery gain	1.6	1.7	1.7	1.8	1.8	1.8	1.9	1.9	0.4
Non-OECD	**2.7**	**2.9**	**3.0**	**3.9**	**4.5**	**5.1**	**5.6**	**6.3**	**2.6**
Natural gas plant liquids	1.6	1.6	1.7	2.1	2.4	2.6	2.9	3.3	2.3
Biofuels[c]	0.4	0.5	0.5	0.7	0.9	1.1	1.2	1.4	3.5
Coal-to-liquids	0.2	0.2	0.2	0.3	0.3	0.3	0.3	0.3	1.9
Gas-to-liquids	0.1	0.1	0.1	0.1	0.1	0.1	0.1	0.1	-0.3
Refinery gain	0.5	0.6	0.6	0.8	0.9	1.0	1.2	1.3	2.7
Total World	**11.6**	**12.3**	**12.8**	**15.0**	**16.0**	**16.9**	**17.8**	**18.4**	**1.4**
Natural Gas Plant Liquids	**8.1**	**8.4**	**8.7**	**10.0**	**10.7**	**11.2**	**11.8**	**12.1**	**1.2**
United States	1.9	2.1	2.2	2.6	2.8	2.8	2.9	2.8	1.0
Russia	0.4	0.4	0.4	0.6	0.7	0.8	1.0	1.2	3.5
Biofuels[c]	**1.2**	**1.3**	**1.4**	**1.7**	**1.9**	**2.2**	**2.3**	**2.5**	**2.1**
Brazil	0.3	0.3	0.3	0.4	0.5	0.6	0.7	0.7	2.4
China	0.0	0.0	0.0	0.1	0.1	0.2	0.2	0.3	8.4
India	0.0	0.0	0.0	0.0	0.0	0.0	0.0	0.0	6.7
United States	0.5	0.6	0.6	0.7	0.7	0.7	0.7	0.7	0.6
Coal-to-liquids	**0.2**	**0.2**	**0.2**	**0.3**	**0.3**	**0.3**	**0.3**	**0.3**	**1.9**
Australia/New Zealand	0.0	0.0	0.0	0.0	0.0	0.0	0.0	0.0	—
China	0.0	0.0	0.0	0.1	0.1	0.1	0.1	0.1	8.6
Germany	0.0	0.0	0.0	0.0	0.0	0.0	0.0	0.0	-0.1
India	0.0	0.0	0.0	0.0	0.0	0.0	0.0	0.0	—
South Africa	0.2	0.2	0.2	0.2	0.2	0.2	0.2	0.2	0.0
United States	0.0	0.0	0.0	0.0	0.0	0.0	0.0	0.0	—
Gas-to-liquids	**0.1**	**0.1**	**0.1**	**0.3**	**0.3**	**0.3**	**0.3**	**0.3**	**5.3**
Qatar	0.0	0.0	0.1	0.2	0.2	0.2	0.2	0.2	11.8
South Africa	0.0	0.0	0.0	0.0	0.0	0.0	0.0	0.0	-0.8
Refinery Gain	**2.2**	**2.3**	**2.4**	**2.6**	**2.7**	**2.9**	**3.1**	**3.3**	**1.1**
United States	1.0	1.1	1.1	1.1	1.1	1.1	1.1	1.1	0.0
China	0.2	0.2	0.2	0.3	0.4	0.5	0.5	0.6	3.0

[a]Other liquid fuels include natural gas plant liquids, biofuels, gas-to-liquids, coal-to-liquids, kerogen, and refinery gain.
[b]OPEC = Organization of the Petroleum Exporting Countries (OPEC-13).
[c]Ethanol values are reported on a gasoline-equivalent basis.
Note: Totals may not equal sum of components due to independent rounding.
Sources: **History:** U.S. Energy Information Administration (EIA), Office of Energy Analysis and Office of Petroleum, Natural Gas & Biofuels Analysis. **Projections:** EIA, Generate World Oil Balance application (2014), run IEO2014_GWOB_LowPriceCase.xlsx.

THIS PAGE INTENTIONALLY LEFT BLANK

www.ingramcontent.com/pod-product-compliance
Lightning Source LLC
Chambersburg PA
CBHW081858170526
45167CB00007B/3064